The Changing Energy Landscape in the Gulf: Strategic Implications

The Gulf Research Center Book Series at Gerlach Press

The GCC in the Global Economy
Ed. by Richard Youngs
ISBN 9783940924018, 2012

Resources Blessed: Diversification and the Gulf Development Model
Ed. by Giacomo Luciani
ISBN 9783940924025, 2012

GCC Financial Markets: The World's New Money Centers
Ed. by Eckart Woertz
ISBN 9783940924032, 2012

National Employment, Migration and Education in the GCC
Ed. by Steffen Hertog
ISBN 9783940924049, 2012

Asia-Gulf Economic Relations in the 21st Century The Local to Global Transformation
Ed. by Tim Niblock with Monica Malik
ISBN 9783940924100, 2013

A New Gulf Security Architecture Prospects and Challenges for an Asian Role
Ed. by Ranjit Gupta, Abubaker Bagader, Talmiz Ahmad and N. Janardhan
ISBN 9783940924360, 2014

Gulf Charities and Islamic Philanthropy in the 'Age of Terror' and Beyond
Ed. by Robert Lacey and Jonathan Benthall
ISBN 9783940924322, 2014

State-Society Relations in the Arab Gulf States
Ed. by Mazhar Al-Zoby and Birol Baskan
ISBN 9783940924384, 2014

Gulf Research Centre Cambridge
Knowledge for All

Political Economy of Energy Reform The Clean Energy-Fossil Fuel Balance in the Gulf
Ed. by Giacomo Luciani and Rabia Ferroukhi
ISBN 9783940924407, 2014

Security Dynamics of East Asia in the Gulf Region
Ed. by Tim Niblock with Yang Guang
ISBN 9783940924483, 2014

Islamic Finance Political Economy, Performance and Risk
Ed. by Mehmet Asutay and Abdullah Turkistani
ISBN 9783940924124, 3 vols set, 2015

Employment and Career Motivation in the Arab Gulf States: The Rentier Mentality Revisited
Ed. by Annika Kropf and Mohamed Ramady
ISBN 9783940924605, 2015

The Changing Energy Landscape in the Gulf: Strategic Implications
Ed. by Gawdat Bahgat
ISBN 9783940924643, 2015

Sustainable Development Challenges in the Arab Sates of the Gulf
Ed. by David Bryde, Yusra Mouzughi and Turki Al Rasheed
ISBN 9783940924629, 2015

The United States and the Gulf: Shifting Pressures, Strategies and Alignments
Ed. by Steven W. Hook and Tim Niblock
ISBN 9783940924667, 2015

Africa and the Gulf Region: Blurred Boundaries and Shifting Ties
Ed. by Rogaia Mustafa Abusharaf and Dale F. Eickelman
ISBN 9783940924704, 2015

Rebuilding Yemen: Political, Economic and Social Challenges
Ed. by Noel Brehony and Saud Al-Sarhan
ISBN 9783940924681, 2015

The Changing Energy Landscape in the Gulf: Strategic Implications

Edited by Gawdat Bahgat

Gerlach Press

First published 2015
by Gerlach Press
Berlin, Germany
www.gerlach-press.de

Cover Design: www.brandnewdesign.de, Hamburg
Printed and bound in Germany by Hubert & Co, Göttingen

British Library Cataloguing in Publication Data.
A catalogue record for this book is available from the British Library.

Bibliographic data available from Deutsche Nationalbibliothek
http://d-nb.info/1070509035

ISBN: 978-3-940924-64-3 (hardcover)
ISBN: 978-3-940924-65-0 (ebook)

Contents

Acknowledgements

Most of the contributors to this volume attended a workshop held in Cambridge, UK in late August 2014. The workshop was sponsored by the Gulf Research Center (GRC). Special thanks go to Dr. Abdulaziz Sager, chairman of the GRC, Dr. Christian Koch, director, Sanya Kapasi, administration manager, and Ms. Elsa Courdier, administrator. Also I would like to thank Kai Henning Gerlach and Dagmar Konrad from Gerlach Press. After the workshop I worked closely with all the contributors to upgrade and revise the papers. I cannot thank them enough for their valuable contributions. Finally would like to thank my assistant Ms. Alexandra Holden for her hard work.

1

Introduction: Energy Security in the Gulf

Gawdat Bahgat

Iran, Iraq, and the six Arab states on the Gulf (Bahrain, Kuwait, Oman, Qatar, Saudi Arabia, and the United Arab Emirates) enjoy several advantages as the world's major oil and natural gas producers. First, together these eight states hold the world's largest proven oil and gas reserves. Their combined shares of the world's proven oil and gas reserves and production are larger than any other region. No region in the world holds as much proven reserves. The high level of production and low (though growing) level of consumption mean that a substantial proportion of the oil produced in this region is exported to the rest of the world. Meanwhile, the massive natural gas reserves and the relatively small volume of production mean that the region has the potential to play a leading role as a natural gas supplier, once gas deposits are fully developed.

Second, the cost of production in the Gulf is one of the lowest in the world. Unlike Russia, the Caspian Basin, the North Sea, and the Gulf of Mexico, most oil and gas fields are either onshore or in the shallow waters of the Gulf. This accessibility means that much of the oil and gas production in the Gulf region is less environmentally challenging and cheaper to produce.

Third, the Gulf region has been producing and exporting oil and gas for decades. Generally the energy infrastructure is well developed. Extensive pipeline networks connect the oil and gas fields to marine export terminals and loading platforms on the Gulf. From there the region has easy access to the high seas and global markets. Unlike other producing regions, shipping ports on the Gulf do not experience major storms or freezing.

Finally, traditionally most of the world's spare capacity of oil is concentrated in the Gulf, particularly Saudi Arabia. This spare capacity serves as an insurance policy against any unexpected interruption of supplies due to natural or political reasons. This concentration of spare capacity can be explained by the fact that oil and gas production is dominated by state-owned national companies. Unlike private international oil companies, which aim mainly to maximizing their profits, these state-owned and state-managed national oil companies are driven by both strategic concerns and commercial interests. Saudi Arabia and other Gulf producers maintain spare capacity to ensure short – and long – term stability of global markets.

Given these advantages and despite rising production in Russia, the Caspian Basin, and Africa, the bulk of the increase in world oil output is projected to come from the Gulf. The oil and gas resources of the Gulf will continue to be critical in meeting the world's growing appetite for energy.

In short, history and geology put Gulf oil and gas producers in the driver's seat. The region has been producing and exporting crude and natural gas for decades and is certain to maintain this policy and status in the future. The projected rise of the Gulf's share in meeting global demand means that major consumers (i.e. China, India, Japan, and Europe) are likely to grow more dependent on energy supplies from the Gulf. The smooth continuation of this mutual dependence between Gulf producers and major consumers requires close cooperation in addressing several strategic and commercial challenges. Some of these challenges are domestic while others are regional and international.

Most of the Gulf governments have achieved a modest success in initiating and implementing economic and political reform. There is much to be desired in pursuing economic development and political liberalization. Equally important, international sanctions, wars, ethnic and sectarian strives, terrorism, and overall regional instability have negatively impacted the full utilization of the region's hydrocarbon resources.

These domestic, regional, and international challenges have raised doubts about the reliability of oil and gas supplies from the Middle East. In the last few decades policy-makers, media outlets, and think-tanks in Washington, Brussels, Beijing, and Tokyo have frequently called for reducing energy dependence on the Middle East. US officials, more than their European and Asian counterparts, have repeatedly talked about "energy independence" and stopping or reducing the nation's "addiction to oil".

This volume argues that such calls are useful for political rhetoric and gaining votes. In reality and based on projections by US, European, and Asian governments as well as by major international organizations, the world will grow more dependent on oil and natural gas supplies from the Gulf. Furthermore, the region's long history of producing and exporting hydrocarbon fuels suggests that concerns over interruption of supplies from the Middle East are exaggerated. A close scrutiny indicates that, with a few exceptions, the region has proven a reliable producer and exporter of oil and natural gas.

For long time the world relied on fossil fuels to meet most of its energy needs. Not only were the prices affordable, but equally important interruption of supplies triggered by political disputes was not an issue. The creation of the Organization of Petroleum Exporting Countries (OPEC) in 1960 as a cartel representing the interest of major oil-producing countries served as a sign that a key change was about to take place. The so-called oil embargo (1973-74) that followed the 1973 Arab-Israeli war represented a turning point in the decades-long perception of energy security. Since then the supplies of oil and natural gas have been interrupted due to political crises. These geo-strategic disputes have also contributed to intense price fluctuation and volatility.

Against this background the interest in alternative energy has emerged and evolved since the mid-1970s. Problems related to safety, reliability, and affordability have slowed down the maturation of nuclear and renewable power. In recent years technological advances and lower costs have convinced many countries to take a fresh look at these alternative energy resources. Furthermore, growing environmental concerns have made energy security inseparable from the transition to a low carbon economy. These developments have broadened the perception and understanding of energy security. In 2011 the International Energy Agency (IEA) adopted a comprehensive approach that includes availability (geological), accessibility (geopolitical), affordability (economic), and acceptability (environmental and social). Finally the availability of reliable supplies at affordable prices with little environmental impact represents only half of the energy equation. The other half is efficient demand. Stated differently, energy security has supply-side and demand-side components.

Within this context, fundamental changes are likely to re-shape the energy landscape in the Middle East in the foreseeable future. Most of these changes have been initiated since the early 2000s or earlier. The main characteristics of this evolving energy landscape include: First, nuclear power still is in infant stage, most countries lack the necessary technical and human infrastructure, while some have the financial resources to launch and fund ambitious nuclear programs, others have harder time allocating these resources. Finally, important issues such as fuel cycle and waste management have yet to be addressed. Second, renewable resources are widely available in the region. However, they are largely under-exploited and under-utilized. In recent years all countries have taken initiatives to diversify their energy mix. The full utilization of renewable resources, however, will take some time and will depend on the building of adequate institutions, enacting appropriate laws and regulations, engaging private sector and encouraging foreign investment. Finally, fossil fuels, particularly oil and petroleum products and to a less extent natural gas, still dominate the energy mix in the Middle East. Their share, however, is likely to decline slowly as nuclear and renewable resources make gains. Subsidies, efficiency, and regional cooperation are likely to impact the evolution of energy strategies in the Middle East and the rest of the world.

The high level of energy consumption in the Middle East and North Africa cannot be divorced from the region's strong tradition of under-pricing energy. In most countries of the region, subsidies for fuels and electricity constitute a significant share of government spending. Indeed, Arab countries, particularly oil producers, are among the largest energy subsidizers in the world. To be sure, most countries in the world subsidize energy in one way or the other.

The argument for energy subsidies in the Middle East (and elsewhere) is based on several environmental, economic, political, and strategic ground. (A) Initiating, developing and marketing renewable and nuclear power usually require government financial support at least in the early stages. All Middle Eastern countries depend heavily on fossil fuels and are among the most polluted in the world. Providing public financial incentives is likely to protect their infant alternative energy programs, mitigate environmental problems, and diversify their energy mixes; (B) Subsidies proved crucial to supply some rural areas with electricity and primary fuels. This development has contributed to economic development, poverty reduction, and job generation; (C) Some of the Middle Eastern countries are critically vulnerable to the fluctuation of oil prices in the global market and its impact on prices in the local economy. Subsidies help to maintain the prices of electricity, natural gas, and petroleum products low and protect against inflation; (D) In major producing countries energy subsidies are part of a broad and un-written social contract between the authority and the people. Under such contract, the authority ensures the supply of energy and other basic commodities at low prices and in return people demand less in terms of political participation and transparency.

Opponents of subsidies argue that artificial low fuel prices create distortive price signals. These low prices are indiscriminate, all energy consumers, rich and poor, benefit from the availability of cheap electricity and petroleum products. But having bigger homes and cars, the wealthy individuals and firms consumer more and benefit more. Furthermore, cheap fuel prices encourage over-consumption and waste. They also create disparity between domestic market and neighboring ones. For a long time heavily subsidized petroleum products wee smuggled to neighboring countries and sold at higher prices. Higher consumption also generates more airborne emissions and greenhouse gases. Finally low prices of oil and gas products leave few incentives to invest and develop alternative energy resources.

This debate over subsidies has intensified in recent years and in response, the authorities in several Middle Eastern countries have considered or started implementing strategies to address resource misallocation and reduce subsidies. New laws and regulations have been enacted to bring petroleum products and natural gas in line with market prices.

This volume examines all of these themes – oil, natural gas, renewables, nuclear power, energy security, and subsidies. Yoshikazu Kobayashi suggests that a country's political power can be broken down into military power, economic power, and power over the opinion. His analysis focuses on economic power in the Gulf. Specifically, he examines the threats to export revenues. These threats are, Kobayashi argues, the shale revolution in North America; the Arab Spring; and the surge in energy consumption. All these threats share a common theme – they will have either few or limited impact in the short-term but will have a more serious and significant impact in the long-term. He strongly urges Gulf countries to invest in energy conservation.

Sujata Ashwarya Cheema's chapter focuses on Saudi Arabia's nuclear energy ambition. She contends that the kingdom's interest in nuclear power is fueled by rising demand for power, population growth, expanding industrial sector led by petrochemical, water desalination, and high demand for air conditioning during the peak summer months. Cheema asserts that Riyadh does not aspire to become a nuclear-weapons state. She argues that the kingdom lacks the necessary natural resources, technological capability, and scientific infrastructure. The country is not known to have any experience in enriching uranium, fuel conversion, fuel fabrication or operating nuclear reactors. Cheema concludes that Saudi Arabia's security depends, to a large extent, on its perception in the United States as a valuable and reliable partner.

Øystein Noreng examines regional energy (and broader economic) relations in the Gulf in light of presumed an agreement between Iran and global powers on the country's nuclear program. He argues that the sanctions regime will not hold much longer and eventually the Islamic Republic will be re-integrated in the global system with or without Washington's consent. Noreng challenges the conventional wisdom by suggesting that the Iranian economy has weathered the sanctions better than many had expected. Tehran's re-integration in the global system, Noreng argues, would contribute to regional political stability and economic prosperity. He takes an optimistic stand and predicts that the antagonism that has characterized relations between Tehran and Riyadh for decades might subside and be replaced by "normal neighborly relations".

Talmiz Ahmad looks at the big picture. He highlights the key developments in the global energy landscape. These include (A) The rise in overall energy consumption and changing shares of each fuel; (B) Technological innovations; (C) Growing concern about climate change and interest in energy efficiency; (D) Shale gas revolution in the United States; and (E) Soaring oil and gas trade between the Gulf states and major Asian markets (i.e. China, India, Japan, and South Korea). Against this background Ahmad provides a detailed analysis of the energy sector in the Gulf and analyzes the challenges the region faces. He argues that a new security paradigm is needed. Such a paradigm would provide a platform for dialogue between Iran and its Arab neighbors on one hand and between Gulf oil and gas producers and Asian consumers on the other hand.

Ali Ahmad and M. V. Ramana examine the growing interest in nuclear power in the Gulf Cooperation Council (GCC) States. The forces driving this interest according to Ahmad and Ramana include reducing heavy dependency on hydrocarbon resources, saving more oil and gas for export, and hedging against Iran's nuclear program. They analyze the potential role of nuclear power and compare the costs associated with it with those associated with alternative sources such as natural gas and solar. Ahmad and Ramana conclude that the cost difference between nuclear and natural gas is dependent on the prices at which natural gas is traded in the international market. At low prices, nuclear power tends to be more costly when compared to natural gas plants.

Dania Thafer focuses on the United Arab Emirates and Qatar. She applies a comparative analysis in examining Abu Dhabi's and Doha's diversification strategies. Specifically Thafer discusses energy subsidies, prices, and consumption. She highlights the need for substantial investment and long-term strategic outlook. In her conclusion Thafer addresses both the demand and supply sides of the energy equation. She urges the authorities in Abu Dhabi and Doha to adapt "a phased-approach" to decreasing subsidies and capitalize on "infinite renewable energy resources".

Mary B. Ayad takes a legal approach in her analysis of the energy sector in the Gulf. She discusses a variety of legal topics including international arbitration, multilateral investment treaty, free-trade agreement, Harmonized International Commercial Arbitration Law Code (HICALC), and Uniform Arab Arbitration Law (UAAL). Ayad examines four types of contracts – administrative contract, production sharing agreements, joint ventures, and service contracts. She reviews the history of oil arbitration in the Gulf, particularly in Saudi Arabia and the UAE. In her conclusion Ayad argues against a regional free trade agreement and for a multilateral investment agreement.

In recent years the International Energy Agency and other major energy organizations have projected a steady rise of the share of renewable energy in the global energy mix. Hanan Albuflasa discusses two scenarios for renewable energy policies in the GCC region. The realization of these two scenarios depends on resources, investment opportunities and technology. Albuflasa highlights the renewable energy projects and the role of both the government and private sector in developing these projects.

Sara Vakhshouri examines the effects of the shale revolution in the United States on the global energy market and its possible implications on the Middle Eastern energy suppliers over the next decade. Still in its early stages, the shale revolution – the combination of computer-aided horizontal drilling and hydraulic fracturing known as "fracking" – is already having a profound multidimensional impact. It is redrawing the United States and global energy landscape, reshaping world energy markets, and beginning to alter global geopolitics. Shale holds promise to substantially enhance US global economic competitiveness and US foreign policy leverage globally. Still, it is worth recalling just how recent a phenomenon this disruptive technology is. There are also continued questions about environmental impact that may limit or even undermine the future of shale gas and tight oil production.

2

Implications of the Emerging Energy Scenario in the Gulf: Towards a Gulf-Asia Strategic Partnership

Talmiz Ahmad

Introduction

The global energy scenario has recently undergone some very significant changes; these include: (a) changes in global energy consumption and the share of different fuels in the global energy mix; (b) application of new technologies that have significantly boosted conventional and unconventional production in different parts of the world; and, (c) increasing concerns relating to energy efficiency and climate change, which have reduced global hydrocarbon demand and encouraged a shift in favour of renewables. Total primary energy consumption is expected to go from 257 mbdoe in 2010 to anywhere between 341-368 mbdoe in 2035, an increase of 30-44 per cent in the next 25-year period. However, energy consumption will display considerable regional variation: Asia, including West Asia, with half the world's population and 60 per cent of global income growth, will account for 70 per cent of world energy consumption growth in 2035.

The production of oil and gas from shale deposits in different parts of the USA, has led to a surge in domestic hydrocarbon production, moving the US from being a major importer towards self-sufficiency, at least over the next 10-15 years. However, over the long-term this will have little effect on the centrality of hydrocarbons and Gulf production in the global energy scenario. In spite of the shale boom, the Gulf still controls 50 per cent of proven global reserves, while North America has 15 per cent and the US just 2 per cent. Cost of production in the Gulf is between $1.5-8 per barrel, while the cost of producing unconventional resources is generally around $80-100 per barrel.

Asian dependence on Gulf imports, presently about 75 per cent, is likely to rise to 90 per cent by 2035; in that year, 90 per cent of Gulf production will also flow to Asian markets. Thus, Asian energy security is crucially dependent on Gulf supplies, giving the Asian consuming nations a direct and abiding interest in Gulf security. The present turbulence in the region, centred around the Saudi-Iran confrontation bilaterally and across West Asia, is therefore a source to legitimate concern to Asian chancelleries.

The paper will provide a detailed analysis of the energy landscape in the Gulf in the global context and will highlight the challenges – economic and political – that this region is facing in maintaining supplies to world markets. After evaluating the present energy and economic ties that bind the Gulf and Asian countries, the paper will explain the need for a new security paradigm in the Gulf, one that would consist of a major initiative by the principal Asian countries – China, Japan, Republic of Korea and India – to promote a new inclusive security arrangement in the region that would provide a platform for dialogue and promotion of confidence-building measures among the principal Gulf countries, particularly Saudi Arabia and Iran, to promote regional security. The challenge before them is obviously quite daunting since, first, the proposal calls for the principal Gulf adversaries to participate in a shared dialogue platform, and, second, it requires the major Asian countries to shed their own differences and develop consensual positions for their common interest. The logic of shared energy security interests should be a powerful motive force to realize this new paradigm.

The Gulf in the Global Energy Scenario

World-wide proven oil reserves are estimated at 1.5 trillion barrels: 73 per cent are in the OPEC countries, with 52 per cent in the Gulf; 14 per cent are in North America, of which 2 per cent are in the USA.[1] Country-wise long-term production is at Table 1.

OPEC and IEA forecasts agree on the central role that Gulf supplies will play from 2020 onwards in meeting global demand. IEA's Chief Economist Fatih Birol has said that Gulf production from now to 2020 will be stable, with the exception of Iraq which will register significant increases. After 2020, however, there would have to be an increase of 6 mbd in Gulf production, for which investments would have to be considered from now itself.[2] Birol has in fact cautioned against suggestions in some quarters that West Asian oil was "not required" due to shale oil production in the US and elsewhere; describing them as "wrong and misleading", he referred to the projected shale oil output as a "surge" rather than a revolution and also one confined to the USA. Birol emphasised: "There is a big Asian demand surge, and the Middle East is the only region that can meet this demand increase. For this, investment is needed in timely manner".

Investments in the energy sector will be crucial to realize the ambitious production capacities required for 2035. The IEA has projected an investment requirement of $37 trillion. It is estimated that West Asia alone may account for investments to 2035 of $3.7 trillion of total energy investment, including investments in nuclear and domestic renewables.[3]

Saudi Arabia will remain a central player in all energy scenarios: it plans to maintain its production capacity at 12.5 mbd, but develop new fields and extend the life of old fields by lowering their production.[4] At the same time, the Kingdom is expected to pursue the use of other energy sources for its electricity production – gas, solar and nuclear, and also unconventional resources such as deep water drilling in the Red Sea and shale gas development. Saudi Arabia proposes to increase domestic gas production from 6 bcf/day at present to 8 bcf/day in 2013 and 10 bcf/day in 2017. It is in this background that OPEC spokespersons and member countries have seen the shale revolution in a positive light, noting that it affirms that, for the foreseeable future, oil will remain "the fuel of choice".[5] Hence, they urge the need for global interest in supporting oil-related technology developments through substantial investments, besides of course promoting price stability by reforming speculative activities which contribute to price volatility.

New Challenges for Gulf Energy

Beyond the hype surrounding the so-called shale revolution, most projections now indicate that shale oil production will be a "brief interregnum" over the next few years, after which, through the 2020s, conventional production from OPEC members, particularly from the Gulf, will once again dominate the global energy scenario. The latter have the advantage not just of abundant supplies but also of the cheapest cost of production: while the cost of production in the Gulf is between $1.5-8 per barrel, the cost of conventional production in other areas goes from around $20/barrel to even beyond $60/barrel. However, even at the higher end, conventional production costs compare favourably with unconventional production: oil sands are between $50-90/barrel; shale oil is $55-100/barrel, while Arctic oil also costs between $40-100/barrel.[6] However, beyond the cost of production, Gulf producing countries face a number of challenges that are examined in the following paragraphs.

The GCC Countries

In normal circumstances, given the high demand in Asia, abundant reserves and low costs of production, GCC producers should look at the future quite confidently. However, this rosy outlook faces some serious challenges which impact on the availability of GCC oil and gas for export. These are: increasing domestic consumption of nationally produced hydrocarbon resources; the grossly inefficient use of these resources at home, and the crippling burden of subsidies on domestic usage. In a Chatham House report on the Gulf energy scenario published in August 2013, the authors made the following observations:

> The systemic waste of natural resources in the Gulf is eroding economic resilience to shocks and increasing security risks. The six Gulf Cooperation Council (GCC) countries now consume more primary energy than the whole of Africa. Yet they have just one-twentieth of that continent's population. Almost 100% of energy is produced from oil and gas without carbon dioxide abatement. If the region's fuel demand were to continue rising as it has over the last decade, it would double by 2024.[7]

The writers went on to point out there were "systemic inefficiencies" in the utilization of energy, with very high energy intensities as compared to other industrialized regions. The situation was further aggravated by major energy subsidy burdens of between 9-28% of government revenues in 2011, more than what was being spent on education and health.[8] In 2011, the six countries of the GCC consumed almost as much oil and gas as Indonesia and Japan combined, and more than the entire primary energy consumption of Africa; the GCC, with just 0.5 per cent of the world's population, consumes 5 per cent of world output.[9] Between 2000-10, GCC primary energy consumption increased by 5 per cent p.a, going from 200 million toe (mtoe) to nearly 380 mtoe; by 2020, this figure could double to 660 mtoe.[10]

Between 2000-09, yearly domestic oil consumption grew by 6.5 per cent; the percentage of oil production that was consumed domestically was a low of 13 per cent in Qatar and a high of 26.5 per cent in Saudi Arabia. In 2009, Saudi Arabia had become the world's number six oil consumer; in 2011, Saudi domestic oil consumption represented lost revenues of $80 billion or 13 per cent of GDP. In Kuwait in 2010, domestic oil consumption in the power sector was about 12 per cent of total production and is expected to 90 to 21 per cent in 2030.[11]

Domestic consumption of national energy resources is being done at prices that are around the cost of production rather than at the international price; this leads to wasteful consumption besides diminishing interest in efficiency initiatives. Electricity is the major consumer of domestic energy production, though the situation varies from country to country: electricity production in the GCC went from 238 terawatt hours (TWH) in 2001 to 462 TWH in 2011, near Brazil (501 TWH) but well above the UK (365 TWH). Most of the electricity is used not in the productive industrial sector but in the residential sector, the share of the latter being: Kuwait 65%; Saudi Arabia 58%; Bahrain 55%; UAE 42%, and Qatar 25%. In 2009, power generation used about a third of total GCC gas production, i.e., 3000 BCF out of a total production of 9500 BCF, which provided 61% of total GCC electricity power generation, with the balance being produced by liquid fuels.[12]

The GCC domestic energy scenario is further distorted by subsidies, which, in per capita terms, are the highest in the world, with the bulk being made up of underpriced electricity; the share of the latter is: Saudi Arabia and Qatar 33%; UAE – 40%; Kuwait 54%.[13] In a recent graph of subsidies published by *MEES*, in 2012 four GCC countries figured in the top 25 countries, in terms of subsidies on fossil fuel consumption:[14]

(a)	Saudi Arabia (no. 2)	:	$62 billion
(b)	UAE (no. 9)	:	$22 billion
(c)	Kuwait (no. 16)	:	$11 billion
(d)	Qatar (no. 20)	:	$8 billion

The situation relating to gas in the GCC is equally parlous. The Gulf countries, including Iran, have a third of the world's gas reserves, with the GCC countries having a share of over 20%. But, actual production in the GCC is quite limited, mainly due to the high cost of production (i.e. $3-9 per MMBtu). Again, pipeline transportation is obstructed by regional political issues, leading Kuwait and the UAE to import gas from Qatar as LNG. In spite of these difficulties, it is anticipated that the GCC will increasingly use gas for its power, desalinated water requirements and for industry. These same projections suggest serious gaps between national demand and domestic production: Bahrain 215 BCF; Kuwait 580 BCF; Oman 1.3 TCF; Saudi Arabia 1.9 TCF, indicating that several GCC countries could become net importers of gas; imports are estimated by EIA at 1.4 TCF in 2025, and double of this in 2035.[15]

This leads to concerns about the actual sources of this imported gas, given the self-imposed moratorium on enhanced production by Qatar (likely to be lifted in 2016) and political difficulties about imports from Iran. There is, therefore, a strong likelihood that much more expensive substitutes will have to be found to sustain domestic requirements, particularly in the power sector. According to recent reports, the UAE imported 17.3 BCM of gas (1.67 billion cfd) in 2012 through the Dolphin pipeline from Qatar, and is investing in increasing the capacity of this pipeline. The UAE is now looking at importing gas as LNG and is building a permanent LNG terminal in Fujairah to handle 1-2 billion cfd of gas from 2016 or 2017, in addition to the floating facility it already has at Jebel Ali. The UAE is also reported to be going further afield to the USA and Canada to import unconventional gas as LNG, besides looking at suppliers in Africa, Asia and the Caspian.[16]

The Saudi gas shortage is so acute that to generate power it was obliged in 2013 to burn 483,000 b/d of crude and 311,000 b/d of fuel oil. According to observers, attempts to switch to gas to generate power are likely to provide only short-term palliatives due to domestic shortages of gas, unless the country imports gas as LNG. The Saudi petrochemicals industry is also looking to the USA to invest in petrochemical projects there due to its inability to meet its requirements of gas from domestic sources.[17] Aramco's annual report for 2013, published in May 2014, announced that with new gas finds, the Kingdom had increased its gas reserves to 288 TCF as against 285 TCF in 2012. The report added that ARAMCO would be using shale gas to fuel a 1000 MW power plant.[18]

GCC countries therefore face a two-pronged dilemma: how to balance rising domestic demand with ever-increasing subsidies and the spending commitments that have emerged after the Arab Spring. The GCC governments' problems are particularly acute since the present domestic energy scenario reduces national revenues both on account of subsidies and through loss of revenues from exports due to rising domestic consumption. The energy analyst Jim Krane has concluded:

> The GCC countries, like numerous oil exporters before them, have encountered the need to prepare their political economies for the plateau and inevitable

> decline of oil exports. The causes of their quandary – and the pressures to overcome it- are internal. Regimes have not been galvanized to seek energy efficiency by an economic shock or international outcry. Rather, energy policy has drifted along on domestic formulae set in the 1970s until becoming apparent that overconsumption is shortening the lifespan of exports and reserves.[19]

On the positive side, over the last five years, there is a better awareness of energy consumption, efficiency and subsidy issues across the GCC: in the UAE, the Abu Dhabi government has set up a task force to reduce electricity demand growth; Dubai is attempting to prepare a comprehensive sustainable energy policy. Saudi Arabia has set up an Energy Efficiency Centre to prepare a master plan on energy consumption, while Qatar is focusing on climate change challenges. Again, all the GCC countries have announced plans to control electricity consumption as also diversification of fuel supply sources for generation. Saudi proposals are the most ambitious: it is looking at obtaining 17 GW of power from nuclear energy, besides the extensive use of renewables (solar, wind, geothermal and waste) by 2032. By 2021, it hopes to obtain a 14 per cent reduction in peak electricity demand and 8 per cent reduction is overall electricity consumption. Above all, by 2021, it plans to bring its energy intensity on par with the G-7 countries. The ARAMCO report for 2013 pledges a reduction by 2.4 mbd in domestic energy demand by 2030 through ongoing efficiency programmes.[20] The Chatham House report suggests that the clean energy and efficiency measures proposed by Saudi Arabia would reduce national oil and gas demand from a projected 4 per cent to 2.8 per cent per annum between now and 2025. Other GCC countries – Kuwait, Qatar, Abu Dhabi, Dubai and Oman – are planning for an extensive use of renewables for power production, with Abu Dhabi and Dubai pursuing nuclear power as well. All of them are also planning extensive reductions of between 20-25 per cent in electricity consumption.[21]

Another area receiving attention in the GCC countries is to promulgate new building codes to promote efficiency, including prescribing better energy efficiency standards for air conditioning systems: codes in Abu Dhabi, Dubai and Qatar now provide for a 70 per cent reduction in energy demand for new buildings and 30-50 per cent savings in existing buildings. Besides energy savings, they are also pursuing water conservation.[22] In May 2014, Saudi Arabia issued a decree enjoining all companies executing projects in the country to adhere to international standards relating to air, water and pollution benchmarks within five years. This is likely to have a major impact on Saudi projects which constitute the bulk of the projects being executed in the Gulf, valued totally at $2.5 trillion. Insistence on green building codes is expected to reduce energy consumption by 40-80 per cent and water usage by 30-40 per cent.[23]

In spite of these positive initiatives, one area that has seen hardly any governmental intervention has been that of energy pricing. The Chatham House report explains it thus:

> ...low energy prices are seen as part of the current social contract between ruling elites and citizens. The popular perception is that, as the countries are major producers of energy resources, low domestic energy prices are regarded almost as a birthright and a mechanism to spread the benefits of these resources to the mass of the population.[24]

This is particularly true of gas prices: local gas prices in Saudi Arabia, UAE, Kuwait and Qatar are about 20 per cent of prices in the USA (i.e., $4 per MBTU), and 40-50 per cent of US prices in Oman and Bahrain; the LNG imported by Kuwait costs even more, $ 13-16/MBtu. Similarly, oil fuel prices across the GCC are between 20-40 per cent of import prices, with only the UAE having prices of 50 per cent of the import price.[25]

The fiscal pressures on the GCC economies have been further aggravated by the Arab Spring, when regimes have felt compelled to radically increase social and economic expenditures to appease different section of local population and neutralise any enthusiasm they might have for political and economic reforms. This has raised break-even prices for oil as the countries struggle to generate the revenues to meet their burgeoning commitments.[26] Break-even oil prices in the GCC in 2012 are at Table 2.

The Chatham House reports ends with calculations of possible savings in energy consumption as a result of implementing technology and efficiency inputs in regard to buildings, air conditioning, transport, and industry. The report concludes that the savings emerging from a basic efficiency scenario would amount to 180 mbtoe per year; at a cost of oil at $100/barrel, this would yield savings of $18 billion per year.[27]

Iraq

In coming years, an important factor in the global energy scenario will be Iraqi production. In July 2013, Iraq published its "Integrated National Energy Strategy" [INES] which sets out three alternative scenarios in regard to oil and gas production [Table 3].

To achieve these targets, the document also estimates investments in the energy sector of the order of $530 billion (at 2011 dollars) between 2013-30. The IEA projections in the three scenarios for 2030 are more modest at: 3.8 mbd, 7.5 mbd and 10 mbd. Even if the "low" scenario of the strategy is obtained, there will be a significant augmentation of global oil supply over the medium to long-term.

However, in spite of these ambitious plans, the present political and economic scenario in Iraq is hardly propitious. The country is today wracked by sectarian conflict, which causes hundreds of casualties every month.[28] Besides the rampant violence, the bureaucratic system is seen as obstructive and corrupt and unable to provide industry and the national community with minimum infrastructure support: just the power sector alone is said to be losing $40 billion per year due to serious shortages. Infrastructure deficiencies are pervasive in the hydrocarbon sector, leading to sharp falls in domestic production and long delays in the contracts being executed by foreign companies. Iraqi production for export is also constrained by lack of storage capacity.[29]

A serious problem that the country faces is the near-independent status of the Kurdish Regional Government (KRG) and its insistence on full control over oil production and sales in that region. In spite of several rounds of discussions, the federal government and the KRG have failed to work out a formula to market KRG oil and share oil revenues. This has restricted both local production and export; thus, the Iraq-Turkey pipeline has been closed since March 2014, while another one million barrels of KRG crude is in storage in the Turkish port of Ceyhan due to the unresolved federal-KRG dispute.[30] The Iraqi political scenario has been further complicated by the emergence of the radical Islamic organization, the Islamic State of Iraq and Syria (ISIS) and its announcement of a "Caliphate" in large parts of northern and north-western Iraq, which have aggravated the country's ethnic and sectarian cleavages and plunged Iraq into a new phase of uncertainty and violence.

In April 2014, there were reports that the INES had actually camouflaged the fact that Iraq was a net energy importer in so far as refined fuels and power were concerned.[31] Analysts believe that the INES document has under-estimated the requirement of natural gas for power, LPG and industrial requirements, and that the total of gas allocated for domestic use would in fact be just sufficient for the power sector alone, and that too only in 2021, and that the additional gas required to meet other projected needs might not be available even in 2025.

At the same time, on the positive side, Iraq continues to award exploration contracts to foreign companies. China is a major role-player: in 2013, it imported about 460,000 b/day of Iraq oil, and has major stakes in the West Qurna-1, Hallaya and Rumaila projects, which are major producers.[32] Present total Iraqi production is about 3 mbd; with additional production from new developments, total production could go up by another 400,000 barrels/day.

Iran

With the US-Iran thaw in September 2013, the interim nuclear agreement in November, and the steady progress since then in Iran's engagement with the P5+1 on the nuclear issue, there is considerable discussion in energy literature on the implications of Iran's return to the global energy scene, unshackled by sanctions.[33]

Before the last crippling sanctions in 2012 that targeted Iran's energy sector and its exports, it used be OPEC's second largest oil producer, at production of 3.7 mbd and exports of 2.53 mbd. (Iran is said to have lost $80 billion in oil revenues due to sanctions.) After the latest sanctions, its production in 2013 was 2.7 mbd, while exports were about 1 mbd; in the first months of 2014, there was a short-term surge in its exports to 1.4 mbd. According to observers, the Iranian oil sector is so run down as a result of sanctions that it would take a few years to reach a production of 3.6 mbd, which would yield exports of about 1.9 mbd. However, the distinguished energy analyst Fereidun Fesharaki is more optimistic: he believes that within ninety days of the easing of sanctions, Iran would be able to produce at the pre-sanction levels of 3.5-3.6 mbd . Energy analyst Robin Mills agrees and believes that only when Iran plans to go beyond the level of production of 3.6 mbd would there be need for considerable external financial technology and human resources.[34]

Irani Oil Minister, Bijan Zanganeh, who took charge in August 2013, is said to have set three targets for his department:[35]

(i) Iranian crude production and exports should be boosted to pre-2012 levels, i.e., to 4 mbd and 2.5 mbd, respectively;

(ii) gas output should be significantly increased with the completion of the development of the remaining 14 phases of the South Pars offshore gas fields; and,

(iii) Iran's petro-chemicals sector should be revived.

Zanganeh is said to be giving immediate priority to enhancing gas production. At present, though Iran is said to have the world's largest gas reserves, it has been running a deficit of 40 BCM/year. Due to gas shortages, Iran's power sector has had to rely increasingly on liquid fuel: while traditionally it depends on liquid fuels for 19-20 per cent of its power generation, in the last financial year (2013-14), its dependence on liquid fuels rose to 48 per cent. Gas shortage has also endangered oil production since increasing quantities are required to be injected into the aging oil fields. According to the EIA, Iran is West Asia's largest user of re-injected gas for enhanced oil recovery operations: in 2010, Iran re-injected more than 1 TCF of gas, about 15 per cent of its total production; in 2020, it is forecast to use between 3.7-7.3 TCF of gas per year for re-injection.[36] Clearly, whatever the quantum of gas used for re-injection, it will have an adverse impact on Iran's exportable surplus. With priority being given to the development of five phases of South Pars, it is expected that by 2017, Iran should have an additional 6.6 bmcf/day of gas, as also 270,000 b/d of condensate. This would add to the 9.7 bcf/day of gas Iran is already producing.

In regard to oil production, Iranian officials have begun to publish very ambitious figures of production, asserting that by 2018 Iranian oil production could reach 5.7 mbd and gas production to 1 bcm/day (35.4 BCF/day).[37] With the easing of political tensions (if not of sanctions), Iran has already begun preliminary interaction with international oil majors, proposing substantial investments across the hydrocarbon value chain in the ongoing five-year economic plan (2011-15): $255 billion for the entire sector, including $155 billion upstream; $62 billion mid-stream and down-stream, and $38 billion for petrochemicals.

Heralding its emerging role in global energy affairs, Iran has entered into three major external energy deals. First, in July 2013, Iran concluded its second gas contract with Iraq to export 850 mcf/day for power generation. Exports are expected to begin from mid-2014, though there have been delays in the completion of the pipeline on the Iraqi side. Second, in August 2013, Iran signed an MOU, valued at about $60 billion, to supply Oman with 880 mcf/day of gas (later said to be 970 mcf/day). According to reports, 70 per cent of the gas would be used by Oman for domestic purposes, while the balance would remain Iran-owned and sold in the world markets after being converted into LNG in Oman. This gas would reach Sohar on the Oman coast through a pipeline measuring 260 km. on land in Iran, with a sub-sea section of 160km. The formal agreement was signed in March 2014 during the visit to Muscat of Iranian President Hasan Rouhani.[38]

The third major agreement is the one Iran is said to be negotiating with Russia. Though details are not available, according to news reports, in January 2014 Iran and Russia were looking at a $20 billion barter deal in terms of which Iran could sell Russia 500,000 b/d of oil in return for goods, service and infrastructure investments of equivalent value. American officials have warned that such a deal could trigger US sanctions against the participating Russian partners, though Russian officials have dismissed these threats. Separately, Iran and Russia are also negotiating power contracts valued at $10 billion to construct new thermal and hydropower plants and transmission networks in Iran, as also exporting 500 MW of power to Iran.[39]

Iran has displayed a new-found confidence in finally terminating the contract of China's national oil company, CNPC, to develop the Azadegan oil field on the ground that the company had conducted "no effective work" since obtaining the contract in 2009.[40] The Iranian oil minister has also put another Chinese company, SINOPEC on notice for making no progress in developing the Yadavaran field. Though China has been Iran's most important energy partner, the Chinese companies, being concerned about exposure to US sanctions, had retained the Iranian contracts without doing much work on the ground. Now, with the cancellation, Iran is conveying a tougher message to its potential partners, while encouraging the return to Iran of former partners such as Total, Shell and Japan's Inpex Corporation. The emergence of ISIS in Iraq, the political uncertainty in Baghdad, and the return of Western armed intervention in the country have generated considerable instability in the region, which is further exacerbated by the uncertainties surrounding Iran's own dialogue with P5+1 on the nuclear issue.

Overview

While there are considerable uncertainties pertaining to unconventional production, both with regard to output within the USA and production in other countries having shale deposits, the long-term scenario relating to Gulf production is much more predictable, particularly since the region has about half the global proven oil reserves of about 1700 thousand million barrels (in 2012). However, Gulf production will also be influenced by certain important factors:

(a) the global economic scenario, particularly the growth prospects of major consumers – China, Japan, Korea and India;

(b) the extent of success achieved globally in applying fuel-efficient technologies, particularly in the transport sector, that would reduce demand for Gulf resources; and,

(c) the application of energy policies world-wide that would take into account climate changes concerns, which would encourage a shift away from hydrocarbons to renewables.

On the supply side too there are a number of uncertainties that will impact on the quantum of Gulf hydrocarbons available for export, or in certain circumstances, even demanded by consumers; these are:

(a) the success achieved by the GCC countries in curtailing domestic demand through the successful application of domestic policies promoting fuel efficiency and climate change interests;
(b) national, regional and extra-regional political developments that will determine the quantum of oil and gas that enters world markets from Iran and Iraq;
(c) the development of new conventional sources in Saharan and sub-Saharan Africa, Central Asia and Latin America, and the continued development of unconventional resources in USA, Brazil and the Arctic, which would impact on demand from the Gulf; and,
(d) price stability at present levels that would encourage further conventional and unconventional exploration and production.

All of these developments will have serious implications for the energy security of the Asian countries. We look at these aspects in the next section.

Gulf – Asia Energy Ties

Growth in developing countries over the last twenty years has lifted millions out of poverty so that today about 525 million people are said to be in the middle class, more than the total population of the EU. Over the next two decades, the size of the global middle class is expected to grow from 1.8 billion at present to 3.2 billion in 2020 and 4.9 billion in 2030, with most of the expansion taking place in Asia.[41] The OPEC Report (2013) projects that 'Developing Asia' will see its share in global GDP rise from 26 per cent in 2010 to 44 per cent in 2035. China will then be the largest economy in the world, while the Indian GDP will nearly be equal to the total European GDP. These changes will be entirely fuelled by the rapidly expanding demand for energy in Asia. Asian demand will have central importance in the long-term global energy scenario:[42]

(i) global energy demand will increase by 33-52 per cent from 2011 to 2035;
(ii) fossil fuels, which accounted for 82 per cent of energy supply in 2010, will constitute

80 per cent of the global total in 2035; each of the three fossil fuel types (oil, coal and gas) will converge around similar shares of about 26-27 per cent; and,

(iii) long-term oil demand will increase by 20 mbd over 2012-35, reaching 108.5 mbd in 2035; of this increase, developing Asia will account for 88 per cent [Table 4].

Long-term energy demand in Asia is at Table 5.

The Asian energy scenario will have the following features:

(i) Asian primary energy demand in the Reference Scenario will go from 4.2 billion toe in 2010 to 7.7 billion toe in 2035.

(ii) China and India will represent nearly 70 per cent of Asian energy demand in 2035, with China accounting for 52 percent and India 18 per cent; Japan's share will decline from 12 per cent in 2010 to 6 per cent in 2035.

(iii) Power generation will be a significant factor in Asian energy use: Asian electricity demand will grow at over 3 per cent per year from 2007 to 2035, with China increasing at 3.25 per cent and India at 5.7 per cent.

(iv) In 2035, Asia will account for 40 per cent of global oil demand, going up from 31 per cent in 2010; Asia will also account for 60 per cent of global oil demand growth. Growth in China's consumption from 2010 to 2040 will be 10.5 mbd while that of India will be 5mbd; after 2035, China will replace the USA as the world's largest consumer of liquid fuels; by 2030, 80 per cent of China's oil will come from West Asia, and 90 per cent in the case of India.

(v) Within West Asia itself, there will be an increase in domestic consumption of 3.2 mbd between 2010 to 2040.[43]

(vi) Asia already absorbs about 75 per cent of GCC energy exports: the extent of dependence on the GCC is as follows: Japan 70%; ROK 60 per cent; India 50 per cent, and China 30 per cent; during the next 10-15 years, even as the US reduces its oil imports, most of the Gulf production will shift to Asia: the Asian giants – China, Japan, Korea and India – will depend on Gulf supplies to the extent of 60-90 per cent, while around 90 per cent of Gulf production will go to Asia.[44]

(vii) Due to technological developments that have significantly increased commercial gas production, and climate changes concerns that have encouraged gas use particularly in the power sector, increase in gas consumption in Asia (including West Asia) will be between 90-110 per cent, as against 20-30 per cent increase in the rest of the world. The Gulf will also play a central role in the global gas trade. It has more than 40 per cent of global proven gas reserves, and is expected to account for 21 per cent of total increase in world production between 2010-40, going 15.9 TCF to 31.5 TCF. Net gas exports from the Gulf are expected to grow at 3 per cent from 2.7 TCF in 2010 to 6.7 TCF in 2040.[45]

Gulf-Asia Economic Ties

In tandem with substantial energy ties existing between the Gulf countries and the rest of Asia, economic ties between the two regions have also been flourishing: non-OECD Asia absorbs 40 per cent of GCC exports. While Japan remains the number one trade partner of the GCC, its share has gone from 23 per cent in 2000 to 15 per cent in 2012. In this period, India has moved from 10th position to the second and obtains 10 per cent of GCC exports. China has moved from a share of 4 per cent of GCC exports in 2000 to 10 per cent in 2012.[46]

Asian countries have also now become major exporters to the GCC: over 35 per cent of GCC imports come from non-Japan Asia, the share of India and China having doubled over the last ten years; in this period, the share of the EU has declined from 30 per cent to 20 per cent. Figures pertaining to GCC's trade with the four major Asian countries are at Table 6.

Just the four major Asian countries accounted for 41 per cent of GCC's two-way trade in 2012, which was valued at about $1500 billion.[47] Exports from Asia consist of machinery and manufactured goods; the break-up of Asian exports to the GCC is at Table 7.

GCC – Asia Investments and Joint Ventures

In 2012, the GDP of the GCC countries was estimated at $ 2.7 trillion, with oil revenues for the year being about $400 billion. By 2030, it is expected that the combined annual oil revenues of the GCC will be $1 trillion. In 2010, the combined national income of the GCC crossed $1 trillion, and is expected to reach $1.5 trillion in 2015 and $2 trillion in 2020. By the end of 2014, the net foreign assets of the GCC countries are expected to rise to $2.3 trillion, while gross financial assets will be $2.8 trillion. The GCC's current account surplus in 2014 will be $287 billion.[48] The GCC countries invest about 80 per cent of their wealth offshore. While investments have traditionally been in Western markets, there is an increasing trend in favour of Asia: cross-border capital flows between the GCC and Asia, which were just $15 billion in 2007, are expected to rise to $290 billion by 2020, when GCC investments in Asia would be about 20 per cent of total overseas flows.

Japan, ROK and China have been remarkably successful in winning major projects in GCC countries in the last few years. In 2007, Japan won the $1.5 billion contract in Kuwait to build a power and desalination plant and major infrastructure development projects in Oman. In 2010-11, Korean companies bagged 47 per cent of all construction contracts awarded in the GCC, and 65 of the major projects. A Korea-led consortium also won the $20 billion contract to build four nuclear power plants in the UAE. Korean firms have projects valued at over $10 billion in Saudi Arabia, which include: petrochemical and refinery facilities, power and desalination plants, and railroads. Korea is said to be moving from economy-oriented exchanges with the GCC to "strategic partnership building".[49]

In an effort to strengthen energy links with China, a number of GCC national oil companies have invested in development of energy-related facilities, including: above-ground storage facilities, refineries and petro-chemical plants. Chinese companies, on their part, are very active in the GCC's construction sector, winning projects worth nearly $10 billion.

Indian successes in the GCC have been relatively modest: the $4.5 billion Oman-India fertilizer project in Sur is the best example of a mutually-beneficial joint venture project, along with the Bharat-Oman refinery project in Bina, India.[50] UAE companies are active in India's infrastructure and real estate sectors developing ports and container terminals, and high-value commercial and residential properties. To encourage GCC investments in India, a number of bilateral funds have been set up such as the India-Oman Joint Investment Fund, which has an initial corpus of $100 million, with a plan to go up to $1.5 billion. The GCC sovereign wealth funds, particularly the Abu Dhabi Investment Authority (ADIA), have invested heavily in Indian equity market on long-term basis.

India has made up for its limited success in regard to executing high value projects in the GCC through the presence of its community in the region: Indians number about seven million in the GCC as a whole, and are the number one expatriate community in every country of the GCC. About 70 per cent are blue collar workers while the rest are professionals and their family members. The Indian community remits to India about $30-35 billion annually.

Ties with Iran

All the four major Asian countries have traditionally been major importers of crude from Iran. However, as a result of US and EU sanctions, imports by all four countries have been drastically reduced, though they will certainly be restored once the sanctions are relaxed. However, the ties of India and China with Iran go beyond energy and have a significance in terms of their strategic interests. In early 2010, the Chinese scholar, Wu Bingbing, had asserted that Chinese strategy in the Gulf rejected the unilateral control of the region by a single power, a clear reference to Chinese opposition to US "hegemony" in the Gulf.[51] Earlier, in June 2009 then president Hu Jintao had said: "Tehran and Beijing should help each other to manage global developments in favour of their nations, otherwise the same people who are the factors of current international problems will again rule the world."[52]

The ties between the two countries are substantial and multifaceted. Besides the energy links, from the early 1980s, China had provided weapons to Iran, valued at approximately $3 billion. In the 1990s, arms exports continued and were valued at over $ one billion. In addition to small arms, Beijing has supplied Tehran with artillery pieces, anti-ship cruise missiles, surface-to-air missiles, fighter jets, tanks, armored personnel carriers, and fast-attack patrol vessels. The value of arms transfers to Iran may have decreased in the current decade. A recent US Congressional Research Service report puts the value of Chinese arms transfers to Iran at $100 million for the period 2002-2005.[53]

While Iran's strategic importance to China emerges from China's principled view that the Gulf should not be dominated by a single hegemonic power that exercises monopoly control over the region's affairs, India's strategic interests are crucial and immediate. They emerge from Iran's position as the "gateway to Central Asia" based on the development of Chabahar port as part of the route that goes to Afghanistan and the road and railway network that is part of the International North-South Corridor (INSTC), which would go up to Russia and beyond into Northern Europe.[54]

The other aspect of India-Iran ties pertains to the shared interests of both countries in regard to Afghan affairs: both of them are wary of Taliban-ISI domination over Afghanistan following the US exit. During his visit to Iran in May 2013, the then Indian external affairs minister Salman Khurshid said: "The convergence of views between India and Iran goes beyond the ambit of bilateral relations and extends to the regional and international arena as well. The Chabahar port project is one such area which reflects our commitment to the stability and peace in Afghanistan:"[55]

There is a little doubt that the future of economic achievement and prosperity in Asia will be crucially linked to the security and stability of the Gulf. In short, the Gulf will be as important to Asia in the 21st century as it was to Europe and the United States in the 20th century, thus providing the principal Asian countries with the rationale for a legitimate role in ensuring that the region remains secure, and the robust and mutually beneficial connectivities in energy, economic and political areas that already exist between Asia and Gulf, are allowed to flourish in the coming decades. The implications of these connectivities and long-term mutual dependence on energy supplies and markets are analysed in the next section.

Towards a Gulf-Asia Strategic Partnership

Asian countries are experiencing a significant change in their status and role in the global economy. Asian economies continued to show fairly substantial growth in output even after the global financial crisis of 2008-10, with China and India showing 9.5% and 7.8% growth respectively, while that of Southeast Asia was 5.3%. Though growth rates in Asia have fallen over the last few years, they are still well above growth in developed countries, with already signs of recovery in several Asian countries.

These trends in favour of Asian economic achievement are expected to continue through the 21st century. By 2025, the Chinese economy will be of the same size as the US economy, with India the fourth largest after Japan; by 2050, the largest economy in the world will be China, which will be twice the size of the US economy, with the Indian economy following a close third, and almost on par with the US. The combined India-China GDP will exceed that of the G-7 OECD economies by 2025; by 2060, it will be more than one and a half times larger. In 2010, India and China accounted for less than half of G-7 GDP, in 2060, the combined GDP of the two will exceed that of the existing OECD area.[56]

Asia's rapid development in the last decade has made the issue of energy security a major challenge for the continent. In Asia's industrialising countries, the average annual energy consumption growth is 3 percent, while this rate is 1.7 percent worldwide. Following the shale revolution in the US and increasing fuel efficiency and use of renewables in the West, there will be sharp declines in dependence on Gulf oil imports in the US and EU. The scenario relating to oil exports from the Gulf in 2035 is expected to be as follows:[57]

- China: 6-8 mbd
- India: 5.0 mbd
- Japan and ROK: 4.5 mbd
- Europe: 2.2 mbd
- US: 0.2 mbd

Hence, over the long-term, the security and stability of the Gulf is of crucial importance to the energy-consuming Asian countries.

A New Regional Security Paradigm

For far too long, the Asian giants have pursued only their economic interests in the Gulf region, while the United States has been responsible for regional security. This it has done by maintaining its political and military hegemony, marked by periodic armed interventions that have wreaked havoc in the countries concerned and encouraged the proliferation of radical elements. It has also stultified the political evolution of the countries in the region, leaving their people frustrated and angry at their marginalization in domestic, regional and world affairs.

Former secretary of state Hillary Clinton set out the link between energy and strategic interests most succinctly when she said: "[Energy] rests on the core of geopolitics, because fundamentally, energy is an issue of wealth and power, which means it can be both a source of conflict and cooperation". [58] In the first decade of this century, in the face of aggressive Chinese initiatives in different parts of the world to participate in E&P contracts, purchase shares in producing fields, and obtain equity in major oil companies in support of its energy security interests, several American commentators had expressed concerns about a "New Cold War", a "New Great Game" and a "New Scramble for Africa". In short, they had viewed Chinese activism in a zero-sum framework, seeing in Chinese successes a threat to US interests.

Though concerns relating to aggressive Chinese activity in the Asia-Pacific continue to agitate some Western and Asian observers, where energy is concerned there is now a greater emphasis on a cooperative rather than a conflictual discourse. The distinguished energy commentator, Daniel Yergin, has explained this thus:

> Virtually all growth in oil demand in the decades ahead is expected to be in these emerging market nations. China and India, as well as Middle-Eastern and other countries, will require substantial new energy supplies to support their rapid economic growth. That necessity creates a great challenge in terms of ensuring that such supplies will be developed in a timely way – and the great risk that, if not properly managed, commercial competition could turn into geopolitical rivalry. Adroit diplomacy will be needed on the part of both developed and developing nations, especially the United States and China. The pressing need for energy locates energy security firmly at the top of the agenda for these two nations.[59]

As a result of the shale revolution, the US will be increasingly self-sufficient in energy, with imports over the long run coming mainly from other American sources. Simultaneously, Asian countries will increasingly depend on hydrocarbon supplies from the Gulf, and hence will have every reason to be deeply interested in the Gulf security scenario. Hence, Asian countries cannot but be deeply concerned about the turbulence and uncertainty prevailing in the Gulf at present as exemplified by the following developments:

- a deep Saudi-Iranian divide, with competitions across West Asia on strategic and sectarian bases;
- the ongoing violence in Syria where the Saudi-Iranian rivalry is engaged in a fierce proxy conflict;
- the emerging US-Iran thaw and the steady progress in the P5+1 nuclear negotiations with Iran, which suggest a possible settlement of the nuclear question, and could open the door for a more influential Iranian role in regional affairs, including in the regional and global energy scenarios;
- Saudi dissatisfaction with US policies in the Gulf and West Asia and deepening estrangement between them; of course, the ideological and geopolitical challenges from ISIS have brought the two together in a cooperative military effort against this new jihadi menace;
- increasing fissures within the GCC itself on the question of ties with Iran and the Muslim Brotherhood;
- continuing conflict in Iraq on sectarian and ethnic basis, which calls into question Iraq's plans relating to increasing its oil and gas production; and,
- above all, the implications of the Arab Spring that challenge the political and economic status quo in all the Gulf Sheikhdoms.

To complicate matters, amidst these disturbing developments that have over-turned the regional security scenario, the US has signaled a shift in its strategic priorities from West

Asia to the Asia-Pacific. This had earlier been projected as a "pivot" to the East, suggesting a diminution in US presence and role in the Gulf and West Asia; more recently, US spokesmen have indicated that this would be more in the nature of a "re-balance" to the East, i.e., a shift of US political focus and military resources to the Asia-Pacific without diluting its presence and role in West Asia.[60]

However, with the principal advocates of the "pivot" to the Asia-Pacific initiative now out of the Obama administration, there are doubts, both in Asia and in the USA, about the extent of the president's commitment to this idea. In fact, it is now widely accepted that what President Obama gives priority to are domestic issues and is strongly averse to any policies that involve a US military role.[61] Hence, not surprisingly, regardless of several assurances from US administration leaders to the GCC countries regarding continued US commitment to their security, the latter remain estranged from the US as it rejects direct military intervention in Syria and continues to pursue the dialogue with Iran.

These developments taken together have important implications for the energy scenario. Thus, the Asian countries that are so crucially dependent on Gulf hydrocarbon resources are now themselves directly exposed to the various risks connected with resource production and transport from West Asia to various Asian destinations, risks that till now were the exclusive concern of the USA as the regional hegemonic power. Given the reluctance of the US to initiate military interventions to sustain its interests, a simplistic solution would be for China, as the major consumer of the region's energy resources, to replace the US to provide security in the Gulf region. This is however not a realistic proposition: first, given its abiding interests in maintaining the security of its regional allies and ensure energy flows, the US is not likely to relinquish its premier military presence in the Gulf. Second, China itself is neither equipped to handle this role, nor has it indicated any interest in raising its profile on its own to include a security dimension.

At the same time, it is important to note that the US, being aware of Asian vulnerabilities on the energy front, is hardly likely to turn it back on its allies in Asia, Japan and Korea, and important political and economic partners like China and India, which are so dependent on Gulf energy. The conclusion that emerges is that the changing geopolitics of energy has created not new areas of competition and conflict but has in fact thrown up new opportunities to pursue shared interests; in short, the need, as Bruce James et al. have put it, is for "a new geopolitical deal over energy and security" that would bring the US, other Western powers and the principal Asian powers in a cooperative security paradigm structured around the GCC countries, Iran and Iraq.[62] The principal Asian energy consumers – China, Japan, Republic of Korea and India – would play a lead catalytic role in pursuing and realizing this new consensual arrangement. Bruce James et al. have captured the logic of the cooperative spirit thus:

> …the future global order needs a much broader grasp of energy issues, and it also needs to have a commitment to low carbon at its heart. By some metrics,

> the U.S. has a declining stake in these discussions, as others become increasingly important consumers to the world's resources. But there are also opportunities for collective action and for cooperation, as rising powers respond to rising levels of risk.[63]

The challenges in realizing this alternative arrangement are daunting for they would require the principal regional role-players to give up their present postures of confrontation and hostility, and engage with erstwhile rivals on the same platform for dialogue, establishment of confidence-building measures, and addressing of issues that divide them in a free and frank environment. Before this, the four principal Asian countries would themselves have to develop the habit of dialogue and development of consensus amongst themselves, a daunting task in itself since Asian countries have little experience of strategic dialogue with each other on Asian issues.

However, the process, through difficult, is not impossible; two useful models are available as guides: the Council for Security and Cooperation in Europe (CSCE), set up on the basis of the Helsinki accords of 1977, that flourished between 1977-94, and brought together contending cold war rivals in constructive dialogue with each other; and the ASEAN Regional Forum (ARF) that is a platform for dialogue on security issues between the ASEAN member countries and countries across the world that have an interest in regional security, and which is in turn supported by a valuable Track-II process, the Council for Security Cooperation in the Asia-Pacific (CSCAP).[64]

The foundations of the proposed new security architecture are solid as it directly addresses concerns relating to regional security. The most urgent issue that impinges on regional security today is the Saudi-Iran divide. While there has been a certain competitiveness between these two Islamic giants since the Islamic revolution, both sides have generally attempted to maintain good working ties, and, on occasion have even had cordial relations. The present discord has its roots in the Arab Spring, particularly its effect in Bahrain, where Saudi Arabia perceived an Iranian role in fomenting the domestic strife. This Saudi-Iran competition has culminated in the mindless blood-letting and destruction that is being witnessed in Syria over the last three years.

In spite of the ferocity of the Syrian conflict and the ongoing vicious war of words across the two sides of the Gulf, sober observers of regional politics do believe that a Saudi-Iran thaw is possible, indeed over-due. The distinguished Beirut-based commentator, Rami Khouri recently pointed out:

> I say there is no real conflict between Saudis and Iranians because these two countries do not threaten each other militarily or strategically, though they do react hysterically when they sense that the other is trying to undermine them ideologically throughout the region.

> Tehran and Riyadh are both regional powers who must be able to protect their national strategic interests in the region. They can best do this by having good bilateral relations, promoting the socio-economic development of all other people in the area, and, together and with smaller partner-states, agreeing on a regional security framework that would be similar to the Helsinki Agreement the United States and USSR worked out in the 1970s.[65]

The cooperative regional security framework proposed above and strongly endorsed by Rami Khouri will have a number of positive features. First, it will promote regional security over the long-term by bringing onto the same platform the principal regional rivals, Saudi Arabia and Iran, whose interaction over time would yield a number of specific and effective confidence-building measures.

Again, it will make it possible for the regional powers to combat the twin scourges of extremist Islam and sectarianism that have proliferated across West Asia over the last three years. Finally, it will enable the Gulf energy producers to consult constructively and pursue policies in their shared interests, a matter of acute urgency in face of challenges from unconventional production; uncertainties relating to supplies from Iraq, Iran and Libya in tandem with the need to maintain the integrity of OPEC as a viable institution; concerns relating to exponential increases in domestic consumption which have led to significant revenue losses; and finally, the need to address fuel efficiency and climate change issues cooperatively.

An added advantage is that the arrangement will take shape incrementally as the level of confidence among the regional players increases; in the interim, existing alliances and security arrangements will remain in place, giving the GCC countries the required level of confidence as they engage with their more powerful neighbours.

The Asian role as a catalyst in shaping the new security paradigm will have the concurrent advantage of promoting broader and deeper connectivities among the Asian producers and consumers that are not possible in a divisive political environment. These could, for instance, take the shape of cross-country and under-sea oil and gas pipelines that would become viable and politically acceptable in the new cooperative atmosphere, perhaps leading in time to the realization of an Asian gas grid that would link the producers of Central and West Asia with consumers in South and East Asia. These possibilities had been envisaged by then Indian Petroleum Minister Mani Shankar Aiyar, in his remarks in Beijing in January 2006, when he had set out a clear vision for Asian resurgence founded on shared energy security interests:

> The Asian quest for Energy Security could lead to Asia regaining its traditional place — a place it has held for thousands of years of recorded history and lost only in the last two hundred years or so – in the vanguard of the advancement of human civilization. The Asian Renaissance brought us all to independence and

liberation. Now, the Asian resurgence depends on energy cooperation in Asia. The 21st century will indeed be the Asian century only if Asian countries — buyers or sellers – join hands together in a continent-wide bid at bringing Asia together and keeping Asia together. I am confident that we will.[66]

Notes

1 Richard G. Newell and Stuart Her, "The Global Energy Outlook", in Jan H. Kalicki and David L. Goldwyn (Ed), *Energy and Security: Strategies for a World in Transition*, (Baltimore: Woodrow Wilson Center Press, Washington DC/Johns Hopkins University Press,, 2013), 54.

2 Alsir Sidahmed, "Fantasy and reality in oil markets", *Arab News*, November 17, 2013, at: http://www.arabnews.com/print/478636 (November 17, 2013).

3 Ali Aissaoui, "Investment for Energy: Looking Beyond Conventional Determinants", *MEES*, November 15, 2013, 24.

4 Guy Chazan, "Saudi faces tricky oil call as shale boom bites", *Financial Times*, December 12, 2012, p. 27; Ed Crooks, "Saudi Arabia welcomes US shale oil boom", *Financial Times*, May 14, 2013, p. 3; "Saudi Arabia to join US as shale gas producer", *Khaleej Times*, October 15, 2013, p. 23; and Saadallah Fathi, "Saudi Arabia: shale options", *Gulf News*, October 21, 2013, B11.

5 Abdalla S. El-Badri, "Middle East and North Africa Energy 2013: The Gulf Economies", Chatham House, London, January 20, 2013.

6 Ajay Makan and Neil Hume, "The Cartel's challenge", *Financial Times*, December 2, 2013, p. 6.

7 Glada Lahn, Paul Stevens and Felix Preston, "Saving Oil and Gas in the Gulf", A Chatham House Report, Chatham House, London, August 2013, vi, at: www.chathamhouse.org/publications/papers/view/193884 (May 10, 2014).

8 Ibid.

9 Ibid, 1.

10 Jim Krane, "Energy Policy in the Gulf Arab States: Shortage and Reform in the World's Storehouse of Energy", (Cambridge, UK: Judge Business School, Cambridge University, 2013), 4, at: www.usaee.org/usaee2012/submissions/onlineproceedings/ (May 14, 2014).

11 Ibid, 6-7.

12 Ibid, 9-12.

13 Ibid, 17.

14 *MEES*, May 2, 2014, 16.

15 Krane, 20-26.

16 *MEES*, January 31, 2014, 8.

17 *MEES*, April 25, 2014, 10.

18 Wael Mahdi, "Saudi gas reserves up as ARAMCO taps new shale deposits", May 15, 2014, at: http://www.bloomberg.com/news/print/2014-05-15/saudi-gas-reserves-up-as-aramco-taps-new-shale-deposits.h (May 18, 2014).

19 Krane, 30.

20 Glada Lahn, p. 11-12; *MEES*, May 16, 2014, 18.

21 Glada Lahn, 16.

22 Ibid, 12-13.

23 DG McCullough, "Saudi Arabia's green decree brings hopes of sustainability", *The Guardian*, May 12, 2014, at: http://www.theguardian.com/sustainable-business/saudi-arabia-green-construction-oil-sustainability-environme (May 14, 2014).

24 Glada Lahn, 21.

25 Ibid.

26 Amy Myers Jaffe and Edward L. Morse, "OPEC: Can the Cartel Survive Another 50 years?", in Kalicki (ed), 130.
27 Glada Lahn, 40-43.
28 *MEES*, May 16, 2014, 16.
29 Ali Merza, "Iraq's National Energy Strategy: Oil and Gas Output, Diversification and Employment", *MEES*, August 9, 2013, 17-19; *MEES*, September 20, 2013, 14.
30 *MEES*, April 18, 2014, 4-5.
31 Ibid, 19.
32 *MEES*, August 30, 2013, 3-4.
33 MEES, May 16, 2014, 20.
34 *MEES*, November 29, 2013, 9-10; *MEES*, January 3, 2014, 8-9; Robin Mills, "Iran's Oil Plans in 2014", January 14, 2014, at: http://www.lobelog.com/irans-oil-plans-in-2014/ (January 15, 2014).
35 *MEES*, January 24, 2014, 4-5.
36 Anthony H. Cordesman, "The Other 'Pivot to Asia' – The Shifting Strategic Importance of Gulf Petroleum", (Washington DC: Center for Strategic & International Studies, November 18, 2013), 6, at: http://csis.org/publication/other-pivot-asia-shifting-strategic-importance-gulf-petroleum (May 5, 2013).
37 *MEES*, April 18, 2014, 6.
38 Ibid; *MEES*, March 14, 2014, 8; Dana El-Baltaji, "Oman Fights Saudi Bid for Gulf Hegemony with Iran Pipeline Plan", April 21, 2014, at: http://www.bloomberg.com/news/articles/2014-04-21/oman-fights-saudi-bid-for-gulf-hegemony-with-iran-pipeline-plan (April 27, 2014).
39 *MEES*, April 11, 2014, 13; "Russia and Iran set to strike $10 bn energy deal", April 28, 2014, at: http:/rt.com/business/155404-russia-iran-energy-deal/ (April 30, 2014).
40 "Iran – CNPC break-up: Tehran eyes the West", *Christian Science Monitor*, May 5, 2014, at: http://monitorglobaloutlook.com/news-story/iran-cnpc-breakup-tehran-eyes-the-west/ (May 8, 2014).
41 Sanjaya Baru, "The Gulf and Asia: Geopolitical implications of Geo-economic Changes". Paper presented at the "Parsing the Reality and Promise of Gulf-Asia Engagement". Seminar, International Institute for Strategic Studies (IISS), Manama, Bahrain. October 15-16, 2011, 2.
42 *WEO 2013* Factsheet; *WOO 2013*, Executive Summary.
43 Cordesman, 2.
44 "Shifting energy markets spur Gulf tilt towards Asia", *Oxford Analytica*, August 27, 2013, at: http://www.oxan.com/analysis/dailybrief/samples/GulfEnergyTilt.aspx (November 18, 2013).
45 Cordesman, 6-8.
46 Kevin Komer and Oliver Maseth, "The GCC going East-Economic ties with developing Asia on the rise", Deutsche Bank Research, Frankfurt, February 18, 2014, 3.
47 Ibid, 4.
48 *Khaleej Times*, Dubai, May 17, 2014, 28.
49 For details of Asia-Gulf economic engagements, see this author's paper titled: " New Silk Roads of the 21st Century: GCC-Asia Economic connectivities and their Political Implications", in Gupta et al., p. 12-25; also see interview with Prof. Tim Niblock at: Abbas Al-Lawati, "The stirrings of a geopolitical shift", *Weekend Review*, *Gulf News*, October 4, 2013, 7.
50 For details of Indian projects in the GCC see this author's paper, "Investments and Joint ventures in India-GCC Economic Ties: "Opportunities and Challenges", in Ranjit Gupta et al. (Ed), *India and the Gulf: What Next?*, Gulf Research Centre Cambridge, 2013, 25-51.
51 Peter Mackenzie, "A Closer Look at China-Iran Relations", Roundtable Report, CAN-China Studies, September 2010, 3.
52 Michael Mazza, "China-Iran Ties: Assessment and Implications for US Policy, April 21, 2011, http://www.irantracker.org/analysis/michael-mazza-china-iran-ties-assessment-and-imlications-us-policy-april-... (4 February, 2013).
53 Ibid.

54 Meena Singh Roy and Ajey Lele, "Engaging Iran in the New Strategic Environment: Opportunities and Challenges for India", *Strategic Analysis*, Vol. 35, No. 1, January 2011, 95-97.

55 Atul Aneja, "India to develop Iranian port", *The Hindu*, May 5, 2013.

56 "Economic outlook, analysis and forecasts: Looking to 2060-Longterm growth prospects for the world", OECD, http://www.oecd.org/eco/outlook/lookingto2060.htm (26 May, 2013).

57 Dr. Mahmoud Vaezi, "Changing Asia: Toward Cooperation and a Common Identity", Centre for Strategic Research, September 2007, http://www.csr.ir/department.aspx?Ing=en&abtid=07&&depid=74&semid=1386 (April 8, 2013).

58 Julia Howald, Stormy-Annika Mildner, Kirsten Westphal, "Tipping the Power Balance? The Shale Revolution and US Foreign Policy", March 2014, p. 16, at: www.transatlanticacademy.org (May 10, 2014).

59 Daniel Yergin, "Energy Security and Markets", in Kalicki, 70.

60 Kristian Coates Ulrichsen, "The Gulf States and the Rebalancing of Regional and Global Order", The James A. Baker III Institute for Public Policy of Rice University, January 8, 2014, 17.

61 M. K. Bhadra Kumar, "Obama resets the 'pivot' to Asia", *Asia Times*, May 9, 2014, at: http://www.atimes.com/atimes/World/WOR-01-090514.html (May 15, 2014); Ulrichsen, 19.

62 Bruce James, David Steven and Emily O'Brien, "Fuelling a New Order? The New Geopolitical and Security Consequences of Energy", Brookings Institution, March 2014, 3.

63 Ibid, 21.

64 For details relating to the implementation of the proposal, see this author's paper, "The Gulf Security Imbroglio: Shaping an Asian Initiative for a New Regional Cooperative Security Architecture", in Tim Niblock with Yang Guang (ed), *Security Dynamics of East Asia in the Gulf Region*, (Berlin: Gerlach Press, 2013), 31-68.

65 Rami G. Khouri, "Changed Ties with Iran will Reconfigure the Middle East", May 17, 2014, 29014, at: rights@agenceglobal.com (May 18, 2014).

66 Talmiz Ahmad, "Geopolitics of West Asian and Central Asian Oil and Gas: Implications for India's energy security", in Ligia Noronha and Anant Sudarshan (eds), *India's Energy Security*, (Oxford, UK: Routledge, 2009), 83.

References

Ahmad, Talmiz, "The Gulf Security Imbroglio: Shaping an Asian Initiative for a New Regional Cooperative Security Architecture". For details relating to the implementation of the proposal, see this author's paper, "The Gulf Security Imbroglio: Shaping an Asian Initiative for a New Regional Cooperative Security Architecture", in Tim Niblock with Yang Guang (ed), Security Dynamics of East Asia in the Gulf Region, (Berlin: Gerlach Press, 2013), 31-68.

Asia/World Energy Outlook", Institute of Energy Economics Japan (IEEJ), January 2013, at: www.eneken.org.jp/data/4683.pdf.

Baru, Sanjaya "The Gulf and Asia: Geopolitical implications of Geo-economic Changes". Paper presented at the "Parsing the Reality and Promise of Gulf-Asia Engagement". Seminar, International Institute for Strategic Studies (IISS), Manama, Bahrain. October 15-16, 2011.

Beck, Martin, "Fracking: A Revolution of the World Energy Market in the Making?', Center for Mellemoststudier, Syddansk University, Denmark, January 2014.

Bhadra Kumar, M. K., "Obama resets the 'pivot' to Asia", *Asia Times*, May 9, 2014, at: http://www.atimes.com/atimes/World/WOR-01-090514.html.

Cordesman, Anthony H., "The Other "Pivot to Asia" – The Shifting Strategic Importance of Gulf Petroleum", Center for Strategic and International Studies, Washington DC, November 18, 2013, at: http://csis.org/publication/other-pivot-asia-shifting-strategic-importance-gulf-petroleum (May 5, 2013).

Dunn, David Hastings and Mark J. L. McClelland, "Shale gas and the revival of American power: debunking decline?", *International Affairs*, 89: 6 (2013), p. 1417-18.

"Economic outlook, analysis and forecasts: Looking to 2060-Longterm growth prospects for the world", OECD, at: http://www.oecd.org/eco/outlook/lookingto2060.htm.

El-Badri, Abdalla S., "Middle East and North Africa Energy 2013: The Gulf Economies", Chatham House, London, January 20, 2013.

Gordon, Deborah. *Understanding Unconventional Oil,* (Washington DC: Carnegie Endowment for International Peace, Washington DC, 2012).

Gupta, Ranjit et al. (Ed), *A New Gulf Security Architecture: Prospects and Challenges for an Asian Role*, (Berlin: Gerlach Press, 2014).

Gupta, Ranjit et al. (Ed), *India and the Gulf: What Next*?, (Cambridge UK: Gulf Research Centre Cambridge, 2013).

Howald, Julia, Stormy-Annika Mildner, Kirsten Westphal, "Tipping the Power Balance? The Shale Revolution and US Foreign Policy", March 2014, at: www.transatlanticacademy.org.

IEC 2013 – Securing tomorrow's energy today: Policy and Revolution Long–Term Energy Security, Deloitte Delhi, February 2013.

James, Bruce, David Steven and Emily O'Brien, "Fuelling a New Order? The New Geopolitical and Security Consequences of Energy", Brookings Institution, March 2014.

Krane, Jim, "Energy Policy in the Gulf Arab States: Shortage and Reform in the World's Storehouse of Energy", Judge Business School, Cambridge University, Cambridge UK, 2013, at: www.usaee.org/usaee2012/submissions/onlineproceedings/ (May 14, 2014).

Kalicki, Jan H. and David L. Goldwyn (Ed), *Energy and Security: Strategies for a World in Transition*, (Washington DC: Woodrow Wilson Center Press, Johns Hopkins University Press, 2013).

Komer, Kevin and Oliver Maseth; "The GCC going East - Economic ties with developing Asia on the rise", Deutsche Bank Research, Frankfurt, February 18, 2014.

Lahn, Glada, Paul Stevens and Felix Preston, "Saving Oil and Gas in the Gulf", A Chatham House Report, Chatham House, London, August 2013, at: www.chathamhouse.org/publications/papers/view/193884 (May 10, 2014).

Maugeri, Leonardo, *Oil: The Next Revolution – The Unprecedented Upsurge of Oil Production Capacity and What It Means for the World,* (Cambridge, MA: Belfer Center for Science and International Affairs, Harvard University, 2012).

Mitchell, John, *US Energy: the New Reality,* (London, UK: Chatham House, 2013).

Mitchell, John with Valerie Marcel and Beth Mitchell, *What Next for the Oil and Gas Industry*?, (London: Chatham House, 2012).

Mitchell, John, *More for Asia: Rebalancing World, Oil and Gas,* (London, UK: Chatham House, 2010).

Noronha, Ligia and Anant Sudarshan (Ed), (Oxford, UK: *India's Energy Security*, Routledge, 2009).

Roy, Meena Singh and Ajey Lele, "Engaging Iran in the New Strategic Environment: Opportunities and Challenges for India", *Strategic Analysis*, Vol. 35, No. 1, January 2011.

Stevens, Paul, *The 'Shale Revolution': Developments and Changes,* Chatham House, London, 2012.

Ulrichsen, Kristian Coates, "The Gulf States and the Rebalancing of Regional and Global Order", The James A. Baker III Institute for Public Policy of Rice University, January 8, 2014.

Vaezi, Dr. Mahmoud, "Changing Asia: Toward Cooperation and a Common Identity", Centre for Strategic Research, Tehran, September 2007, http://www.csr.ir/department.aspx?Ing=en&abtid=07&&depid=74&semid=1386.

World Energy Outlook [WEO] 2013 Factsheet: How will global markets evolve to 2013?, IEA, November 12, 2013, at: www.iea.org/newsroomandevents/.

World Oil Outlook – 2012, at: http://www.opec.org/opec_web/static_files_project/media/downloads/publications/WOO2012.pdf.

Yergin, Daniel, *The Quest – Energy, Security, and Remaking of the Modern World,* (London, UK: Allen Lane – an imprint of Penguin Books, 2011).

Appendix

Table 1 **Country-wise Oil Production (unconventional production in brackets) [in mbd]**

Country/Region	2010	2035
USA	9.7 (0.9)	11-15 (2-3)
Canada	3.4 (1.5)	5-7 (3-5)
Total North America	**16.1 (2.4)**	**17-24 (5-8)**
GCC/Gulf		
Saudi Arabia	10.5	11-15
UAE	2.8	2-3
Kuwait	2.5	3-4
Qatar	1.4	2-3
Total GCC (excluding Oman)	**17.2**	**18-25**
Iran	4.3	3-4
Iraq	2.4	4-8
Total Gulf	**23.9**	**25-37**
Total World	87 (4.2)	100-115 (8-18)

Source: Energy and Security: Strategies for a World in Transition, p. 56.

Table 2 **Break-even Oil Prices to Meet Fiscal Commitments [in $ per barrel]**

GCC Country	2008	2012
Bahrain	75	119
UAE	23	84
Saudi Arabia	49	71
Kuwait	33	44

Source: Energy and Security: Strategies for a World in Transition; p. 130.

Table 3 Present and Long-term Iraqi Oil Production Outlook (mbd)

Year	Low	Medium	High
2012	2.3	-	-
2017	5	6.5	13
2020	6	9	13
2030	8	9	13

Source: MEES, August 9, 2013, p. 17-19.

Table 4 Long-term Region-wise Oil Demand Outlook (in mbd)

Region/Country	2012	2020	2035
OECD	46	44.2	40.4
Developing Countries	37.8	46.6	62.1
India	3.7	5.0	9.3
China	9.7	12.7	17.5
Eurasia	5.0	5.5	6.0
World	88.9	96.3	108.5

Source: World Oil Outlook 2013, Executive Summary, OPEC, November 7, 2013, p. 10.

Table 5 Asian Present and Long-term Energy Demand

Energy Source	2009 [Mtoe]	Share (%)	2035 [Mtoe]	Share (%)
Coal	2165	47	3101	41
Oil	1132	25	1687	22
Gas	437	10	1063	14
Nuclear	146	3	522	7
Renewables (Inclusive of biomass & waste)	609	13	1049	14
Hydropower	84	2	201	3

Source: IEA World Energy Outlook 2011, p. 570-90.

Table 6 Trade Figures (2012) [in billions of $]

Country	GCC exports	%	GCC imports	%	Total 2 way Trade
Japan	143	15.3	27	5.6	170
India	98	10.5	55	11.3	153
ROK	95	10.2	22	4.4	117
China	97	9.8	60	12.3	152
	428	**46%**	**164**	**33.6**	**592**
Total	930		497		

Source: "The GCC going East", Deutsche Bank Research, p. 4.

Table 7 Category-wise share of Asian exports to the GCC

Item	% share
Manufactured goods	32.6
Machinery and transport equipment	15.1
Jewellery	8.6
Telecommunications	7.4
Food stuffs	7.0
Refined petroleum	5.0
Plastics and chemicals	2.5
Others	6.2

Source: "The GCC going East", Deutsche Bank Research, p. 5.

3

Political Détente and Economic Tensions

Øystein Noreng

Iran's Comeback – a Game Changer

The key assumption of this paper is that, as it appears at the time of writing, spring 2014, a nuclear deal permitting the reintegration of Iran into the world economy is likely, notwithstanding opposition among hardliners in Iran as well as in the US Congress. An understanding on the nuclear issue would reduce political tensions not only between Iran and the West, but also with the states of the Gulf, first of all with Saudi Arabia.[1] For many years, debate on Iran has focused on nuclear politics, economic aspects have been largely ignored. The narrow economic discussion has highlighted oil and natural gas, ignoring the industrial and manufacturing potential. Indeed, on the condition of being readmitted into the world economy with sanctions lifted, Iran has a potential for high economic growth. If, by chance, negotiations should stall or an understanding with Iran should be blocked by the US Congress, sanctions reasonably would grind down gradually, possibly on a country-by-country basis, leaving the United States and a few other countries isolated. In that case, Iran's comeback would be slower, but hardly halted. A major difference is that in the first case, Iran would reintegrate the world economy with the approval of the United States; alternatively, it would reintegrate the world economy in opposition to the United States, most likely with close ties to China and Russia.

The Iranian economy has weathered the sanctions better than many expected.[2] Indeed, sanctions have spurred the country to find new ways, accelerating economic reform. Even if Iran still has huge liabilities in its political economy in the form of cronyism, corruption, barriers to entry, imperfect competition, and oligarchic structures in control of industry and key services, the assets are also important. Iran has a large, relatively educated young population, meaning a skilled workforce as well as a big home market. The infrastructure has been developed through substantial investment in roads, railways, telecommunications and power supply. Local governments are able to provide basic education and health services. Consequently, Iran is ripe for foreign direct investment; it should be open for business, possibly on a large scale.

Against this backdrop, reasonably, lifting sanctions would lead to economic growth, not only based on oil and natural gas, but also in manufacturing. Iran would gain in economic weight in the Gulf, in addition to its demographic weight. Saudi Arabia seems to have realised the change, renewing relations with Iran. The antagonism that for almost a decade has marked relations between the two, seems to subside. It might be replaced by more normal neighbourly relations, at least for some time.

Recurrent Economic Tensions and Political Fallout

Political détente in the Gulf makes Iran appear even less as a military threat to the Gulf States. Instead, the reintegration of Iran into the world economy carries a risk of unleashing economic tensions in the Gulf, especially over oil. Iran's demographic and economic weight provides a basis for political influence in the Gulf region, most likely to be bolstered by close ties with Iraq, with repercussions for the oil market. Historically, tension over oil market shares and prices has marked relations between Iran, Iraq, Kuwait, Saudi Arabia and the UAE. In relation to oil market strategy there is a structural divergence of interest, with populous Iran and Iraq wishing both high prices and high market shares for themselves, eventually at the expense of the less populous GCC states that generally have had an interest in lower oil prices, but also high market shares.

Conflict over oil prices and market shares was one of the factors driving Iraq in 1990 to occupy Kuwait. The breakdown of the OPEC consensus in 2011, with Saudi Arabia unilaterally expanding volumes, indicates that such conflicts are still real. This was corroborated by disagreement over the succession of the OPEC Secretary General in 2012. Even if Iran and Iraq have different and competing interests, they also have salient common interests in relation to the oil market and thus to Saudi Arabia and the other Gulf states. With a nuclear deal, Iran is likely to open up for foreign oil investors, most probably leading to a considerable increase in oil extraction capacity. Likewise, in Iraq, volume growth is essentially a matter of time. Consequently, the outlook is for a considerable volume growth from the two most populous countries around the Gulf.

The advance of unconventional crude extraction and the erosion of the predominant position of oil in the transportation sector exacerbate the dilemma, as new competition sets new limits to oil pricing. More than before, rising prices risks compromising oil demand, whereas lower prices may undercut competition and secure long-term volume and income growth, but at the expense of short-term earnings. Against this backdrop, the issue is to what extent total oil supplies from the Gulf will expand or to what extent Saudi Arabia and the other GCC members will renounce on market shares and earnings in order to sustain oil prices and accommodate Iran and Iraq.

Iran's demographic weight can sustain military power as well as economic clout. Iran has become Iraq's major trade partner in addition to developing close trade links with Turkey. Iraq apparently aims at once more to become an important military power in the region. Rising oil exports enhances economic importance. Fending off Kurdish attempts at controlling oil revenues is an attempt to strengthen the Baghdad government. Consequently, until the summer of 2014, political weight in the Gulf region seemed to shift from the GCC states to Iran and Iraq. By the summer of 2014, the precarious cohesion of Iraq evidently weakens the country's regional influence. A fragmentation of Iraq would strengthen Iran's position in the Gulf, insofar as a Shia-majority rump state in southern Iraq would be more dependent on Iran for trade and security. In case Iraq should survive as a political entity, the combination of a large population and rising oil revenues should underpin political clout.

The prospect of rising Iranian and possibly Iraqi influence has provoked different reactions on the southern shore of the Gulf. So far, Saudi Arabia seems bent on resisting the rise of Iran, by supporting opposition to the Syrian regime and by trying to corral the smaller Gulf States into a union, possibly a confederation with a common currency. This initiative has been resisted by the states concerned; they seem willing to accommodate to a more influential Iran.

Consequently, until the summer of 2014 the stage seemed set for an isolation of Saudi Arabia, at least temporarily, unless the Kingdom also would accommodate to the new situation and a resurgent Iran. Insofar as Saudi Arabia considers closer ties between Iran and Iraq as a security risk, it has incentives to support the Sunni Iraqi opposition and eventually Kurdish independence, in addition to the Sunni forces in Syria. The civil war in Syria can to some extent be seen as a fight by proxies between Iran and Saudi Arabia. Likewise, Saudi support for Sunni insurgents in Iraq could also be seen as a fight against Iranian influence in Baghdad.

The sudden emergence of IS in the summer of 2014 represents a major threat to all established governments in the region. A common enemy might even bring Iran and Saudi Arabia closer together, in spite of a historical record of rivalry and distrust. The IS in Iraq and Syria is indeed a game-changer. It might represent a more acute danger to the Saudi regime than to Iran. Against that backdrop, the IS might provoke some understanding between Iran and Saudi Arabia on the need to stabilise Iraq and Syria. The practical implementation of such an understanding would, of course, require politicians in Iraq and Syria willing and able to bridge the cleavages and reconcile the parties to the civil wars raging.

Correspondingly, insofar as the Iraqi government considers Saudi support for Iraqi Sunni forces as a security risk, it has reasons to align itself more closely with Iran. Finally, if Iran were to see Saudi support for Sunni forces in Iraq as a security risk, it would have incentives to draw Iraq closer. Against this backdrop, there is a risk of escalating conflict, which could also include Iraq. As a countermeasure to Iran's rising influence, Saudi Arabia has the ability to raise oil output in order to bring prices down and compromise Iran's export earnings. Such a move would eventually raise tension in the Gulf, with the prospect of further isolating Saudi Arabia.

The alternative is to get to terms with Iran, which may be happening in 2014, as mentioned above under the impact of IS. The replacement of Prince Bandar as head of

Saudi intelligence is but one signal. The Saudi government has expressed the intention to clamp down on Saudi citizens fighting among the Sunni fundamentalist opposition in Iraq and Syria. The question is what measures are really being taken, given the number of Saudis who have joined the jihadist forces in Iraq and Syria. Halting the flow of money is the most urgent task, but it is uncertain if any measures are being taken to stop the flow of money from Saudi Arabia and the other Gulf States to the jihadists, first of all IS.

Oil and Regional Conflict

Oil is a regional issue. To the extent that Saudi Arabia and other Gulf countries consider Iran a security threat, they have incentives to weaken Iran through lower oil prices by raising export volumes. And if Iran considers Saudi Arabia and other Gulf states an economic threat through their ability to lower oil prices, that gives it incentives to make an alliance with Iraq, perhaps to threaten the Gulf countries. Against this backdrop, prospects for the oil market and oil prices are conflicting, possibly indicating instability and discontinuities. There are good reasons to expect higher oil volumes from the Middle East to the market—not only from Iraq as its industry is being restored, but also (with political motives) from Saudi Arabia and other Gulf countries in order to hit Iranian oil revenues through lower prices. There are equally good reasons to expect Iranian countermoves, perhaps together with Iraq, if that country remains a political entity, aiming at higher oil prices.

On oil policy, Saudi Arabia and the other rich Gulf States face a dilemma. On one hand, they need revenues to finance rising budget needs, but on the other hand they need to secure the long-term competitiveness of oil and especially their own oil facing rising supplies of both conventional and unconventional oil from other parts of the world. The critical case is Saudi Arabia with the largest reserves and extraction capacity. In 2013, the kingdom had a current account surplus of US$ 130 billion, the difference between the value of exports at US$ 377 billion and imports at 247 billion. Petroleum exports accounted for US$ 322 billion, averaging 7.57 million barrels a day, mbd, of which exports of petroleum products averaged 0.8 mbd. Thus, with other export revenues of US$ 55 billion, the kingdom would require US$ 192 billion in oil revenues to cover imports at present levels. The figure translates into a daily oil income need of about US$ 530 million. This gives Saudi Arabia a considerable leeway in petroleum policy. At the present (autumn 2014) price of US$ 100/bl., the kingdom could eventually reach that target by exporting just 5.3 mbd, much less than the current level. Alternatively, keeping the current export level, Saudi Arabia could reach the above mentioned income target by an oil price of US$ 69/bl. Moreover, if it so wanted, the kingdom could expand output to 10 mbd. With a price of US$ 53/bl., the income target would still be reached.

Consequently, Saudi Arabia, together with Kuwait and the UAE, has a real option of expanding output and lowering oil prices to defend the market share for its own oil and for conventional oil in general. Moreover, the above discussion only focuses on annual balances, abstracting the accumulation of surpluses over many years, providing financial resources for oil policy flexibility. Indeed, from 2004 to 2013, Saudi Arabia's cumulative current account surplus amounts to about US$ 1 trillion.

Thus, there is no question that Saudi Arabia and some other Gulf monarchies have the financial muscle to shake up the oil market. The question is rather what interest they might have in doing so. Spare capacity provides an opportunity to offset lower prices by higher volumes. Historically, given the properties of oil demand, oil exporters cooperating in the market have shown a preference for making more money with less volume. The notable exception is the mid-1980s, when oil demand was stagnant and non-OPEC supplies were rising. With US shale oil, Mexico reforming, new discoveries and investment in Africa and in high-cost prospects offshore and in the Arctic, within a context of stagnant oil demand, a similar situation seems to emerge by 2015; supply seems set to increase more than demand at least for some years. That is a recipe for lower prices to clear the market unless the core producers withhold volumes and renounce on market share.

The advantage for Saudi Arabia and the other Gulf States would be a higher market share for themselves as well as for conventional oil in general, compromising the economics of shale oil and gas and costly deep-water and Arctic prospects. The world economy would benefit by generally lower energy costs. Trade balances of oil importers in southern Europe and Asia would strengthen. Most likely, Saudi Arabia would consider a threshold of pain for the US shale industry, but the kingdom might have a political interest in keeping a US dependence on Saudi crude. Price is the appropriate mechanism. Lower oil prices would hit Russia and Iran, which might not be bad from a Saudi perspective.

The additional supply potential in the Middle East is not limited to Saudi Arabia and the Gulf States. In an extreme case, a split-up of Iraq with an independent Kurdish state under Turkish protection might release large volumes of oil from present-day Iraqi Kurdistan, through Turkey and onwards to the Mediterranean market. At the same time, the revenue loss for the remaining part of Iraq might spur stronger policies for volume growth, despite the organizational and administrative problems. Consequently, a partition of Iraq might bring higher oil supplies and lower oil prices – unless it should lead to a regional crisis and perhaps war that would interrupt oil supplies.

After a Century

The Middle East state configuration from the settlement of the First World War is under severe stress after a century. Moreover, the balance of foreign power interests in the region is also changing. Oil is a key factor, in the early 21st century as one hundred years earlier. A few years before the First World War, a joint British–Dutch–German oil company was set up to explore for oil in the then Turkish Iraq. British oil interests were decisive for the dismemberment of the Ottoman Empire, the ensuing political fragmentation of the Arab lands and the suppression of Kurdish aspirations for independence.[3]

British and French promises during the First World War to entice cooperation in the fight against the Ottomans were new Arab and Kurdish states. One outcome could have been an independent Arab nation-state, covering the Fertile Crescent, comprising Lebanon, Syria and Iraq, in addition to present Jordan and Hedjaz, with ports on the Red Sea, the Mediterranean and the Gulf. Eventually, this would have made a country with a large population and ample resources, including agricultural potential and much oil. It would have been ethnically and religiously mixed, but that had been the tradition in the Middle East. It is questionable whether Kuwait and Saudi Arabia without Hedjaz would have been able to stay outside such a large Arab nation-state. Such a consolidated Arab political entity was contrary to the British strategy of fragmentation.[4] Likewise, a Kurdish state comprising areas of present-day Iraq, Syria and Turkey, would also have had oil, in Mosul, and a large population able to sustain a strong state.

Control of oil was the British motive, not ethnic homogeneity for establishing Iraq. This priority contrasts with the alleged tenets of the Versailles Treaty, indicating that Britain and France, with the tacit support of the United States, openly did not practice in the Middle East what they preached in Europe. Since then, political legitimacy has always been a serious problem in Iraq. Among the spoils of victory, oil in Mesopotamia was the prize, especially in the Kurdish-speaking province of Mosul.[5] At a late stage during the First World War, the area was transferred from the French sphere of interest to the British one, in return for a British recognition of a French stake in Mesopotamian oil.[6] The British joined the Kurdish-speaking Mosul province with the Arabic-speaking Basra and Baghdad provinces to make the new state of Iraq. The majority of the Arab Iraqi population is Shia, but the British masters transferred political power to the Sunni minority.[7] It could be argued that at the time, the divide between Sunni and Shia was less acute and had less political resonance than today. Nevertheless, the divide at that time had a social significance, as the British masters relied on essentially Sunni landowners, whereas the poorer parts of the population was Shia. Even if there was limited sensitivity to democratic norms, British supported Sunni rule was disturbed by Shia and Kurdish insurgencies.

Generational Conflict and Political Change

A century later, the legacy of the Sykes-Picot agreement is crumbling, potentially with important repercussions for regional politics as well as the oil market. An explicit aim of the IS is to overcome the Sykes-Picot legacy by establishing a new state encompassing parts of both Iraq and Syria. Since 2011, a wave of protests spurred by generational conflicts and economic grievances has been shaking the Arab world regardless of the regimes in place. After a few years, the movement seems to have stalled, except in Tunisia, as witnessed by the setbacks in Egypt that has a military dictatorship, again. The underlying problems that caused the protest movement have not been solved. The main driver, generational change, simply cannot go away.[8]

The Arab Spring has been conspicuously different from earlier political mass movements in the region by the call for freedom and democracy, and a liberal state, and implicitly a secular society where religion is a private matter. Previous mass movements in the Arab world have been more nationalist than democratic. Indeed, the democratic and anti-authoritarian character of the Arab Spring is underlined by its anti-hegemony. From this perspective, the Arab Spring seems to mirror the 19th century liberal revolutions in Europe rather than the 20th century socialist ones. The emphasis on a vision of freedom and democracy distinguishes the Arab Spring as the movement of young, educated people, contrasting with the Moslem Brotherhood that appears as literally older, less educated and less visionary, and to some extent part of the establishment.

The driving forces of the Arab Spring, essentially the generational conflict and the demand for economic and social progress, meaning employment, income and welfare, will divert more re-sources to civilian tasks, leaving less money for military forces and arms procurement. Moreover, due to generational dynamics, the Arab Spring cannot be rolled back. It may be delayed, but it is likely to affect the entire region, with no exemption for the conservative monarchies. The question is the timing and the sequence of events.

Consequently, civilian economic issues are increasingly important in Middle Eastern and Gulf politics, within the various countries as in relations between them. Even if economic disputes, especially over oil, carry a potential for conflict, economic concerns represent a check on the use of military force. Put simply, countries that may have economic reasons to wage war often lack the economic resources to do so. Ever more, the Middle East and the Gulf reveal fault lines and cleavages that cannot be overcome by military force, but which require a political solution. Therefore, the Arab Spring has the potential to limit military budgets and enhance power sharing in the region. Otherwise, economic stagnation and social grievances are likely to build up pressure for more violent political change. Egypt is a case in point. The issue is what civilian pressures might push policy makers in the direction of reform within what time frame.

In the longer run, the movement has a potential to modify the state configuration of the Middle East and to rebalance foreign relations, with Iraq remaining the most precarious case. In hindsight, the substitute for one Arab nation state has been several Arab states. Their number, diversity and conflict potential serve foreign oil interests. Fragmentation of Arab politics provides an easy playing field for foreign intrigue. The Arab Spring changes politics, within and between states. With largely identical driving forces, political conflict in the Arab world is increasingly within the various countries rather than between them. Eventually, in the longer run, this might contribute to a stronger common Arab identity and more cooperation among the various states. Such an outcome would facilitate Arab cooperation on oil matters. Most importantly, as discussed above, the integrity of Iraq is under pressure, directly linked to the rise of Kurdish nationalism. That is but one of the critical points with a potential for triggering major changes in the Middle East, also affecting the oil market.

The Reintegration of Iran, Repercussions for Syria and Turkey

A second critical point is the reintegration of Iran into the Middle East mainstream, and especially the Gulf. Even if oil exports would increase with the lifting of sanctions, Iran is in need of industrial investment to create jobs. This not only requires capital, but also reforms. Currently, the political economy of Iran is dominated by powerful vested interests, rival oligarchies for whom the formal distinction between the private and the public sector has little real meaning. Sustainable economic development in Iran is hampered by administrative incompetence, corruption, barriers to entry, defective competition, as well as wanting transparency and accountability.[9] Tellingly, the wide-ranging agreement with China in 2004 to facilitate industrial investment has had little practical effect, essentially due to administrative incompetence on the Iranian side. Consequently, petroleum revenues that require a limited effort and that be fought over, do not challenge the vested interests. A policy for large-scale direct investment in manufacturing would require significant reforms that would challenge the power structure. From this perspective, the reintegration of Iran into the world economy would depend on substantial domestic change. From this perspective, a lifting of sanctions on Iran would delegitimise many of the repressive policies of the present regime. Consequently, barring a nuclear agreement risks obstructing political change in Iran.

A third critical point is the integrity and stability of Syria. By its ethnic and religious diversity, Syria appears as a microcosm of the Middle East. So far it has been held together by force, since 1970 by a government with its base in Alawite minority, but enjoying considerable support from other groups. All six neighbours are affected by the unrest and the civil war in Syria, and all six neighbours are actively engaged in the country. Any disintegration of the Syrian state would involve each neighbour. Although Syria is not a large oil exporter, historically it has been an important transit country for pipelines.

A fourth critical point is Turkey's re-entry in the Middle East. After almost one hundred years, Turkey has made a remarkable comeback, especially in Iraq. Since the defeat in the First World War, Turkey has been conspicuously absent from Middle East politics. Turkey has traded with Middle Eastern neighbours, but has hardly been involved politically. Indeed, so far, Turkey emerges as the commercial winner of the Iraq War. As a major importer, Turkey has significant oil and natural gas interests in the region, first of all in Iraqi Kurdistan.

Both Iran and Turkey claim credits for the Arab Spring. Iran claims to be the inspiration for the overthrow of pro-Western authoritarian regimes having ousted the shah decades ago. Turkey pretends to be the model of Islamic democracy or democratic Islam. The Arab spring also affects relations between Iran and Turkey. In the past, the two countries have been competing for influence in Iraq and Syria. Civil wars and the emergence of IS have brought Iran and Turkey closer together. Security concerns foster a congruence of interests and the two countries are now actively engaged with each other.

Iran, a Shiite country and an Islamic republic with serious economic and political problems, does represent a distinctive model for Middle East societies, but it is hardly attractive except in fairly limited circles. Contrary to its own hopes and other countries' fears, Iran has not been able to export its Islamic revolution. Over time, the oppressive features of the Iranian regime have proved to be drawback in the quest for influence in the Middle East. Iran is no ideal for the Arab Spring. On the contrary, many of the key Arab Spring issues are also pertinent to Iran. Iran was contented by the Tunisian revolution and greeted the overthrow of the Mubarak regime in Egypt, only to be disappointed by Egypt's reticence to engage in closer relations, although Egypt did open the Suez Canal for Iranian naval vessels. Since the 2013 coup d'état in Egypt, relations are sour. Instead, as mentioned, under President Rouhani, Iran's position is improving in the Gulf. Moreover, there seems to be an even stronger pressure from the younger generation, with no memory of the shah or the war with Iraq, for both jobs and cultural, social and political change.[10]

By contrast, Turkey, the Sunni Muslim country, has applied secularism since early 20th century after the fall of the Ottoman Empire. The Turkish Justice and Development Party (AKP), a liberal conservative party with Islamic flavour, was founded in 2001. It gained more power and won the 2002 election. Since then, Turkey has enjoyed remarkable economic growth and social progress, appearing as a more attractive model for development, enhancing Turkish influence in the Middle East. Consequently, Turkey has been seen as a relevant long-distance ideal for the Arab Spring. The authoritarian twist of the Erdogan government has, however, done much to tarnish Turkey's image in the Arab world. Erdogan's policy on Syria has been unsuccessful, siding with insurgents against an Assad government the surprisingly seems to survive. Turkey's policy on Iraq and Kurdistan is incoherent, driven by a desire to get preferential access to oil and natural gas in Northern Iraq, but tempered by a fear of alienating Baghdad. Consequently, Turkey's influence in the Gulf, and especially in Iraq, that has been strengthening, is again weakening. The increasingly authoritarian character of the Erdogan regime is a liability at home as well as abroad.

To sum up, the generation-driven quest for political change affects the entire Middle East, also the Gulf region. It is picking up in Iran and Turkey. In the Arab world, the process is proceeding in Tunisia, it has stalled in Egypt, but most probably only for some. The elements are certainly present in the Gulf, but local circumstances and the oil wealth will temper the further process.

Oil is Still the Issue

For the rest of the world, the salience of the Middle East and North Africa is based on the fact that oil is indispensable to the modern economy due to its physical properties.[11] Crude oil and oil products such as gasoline, diesel and fuel oil are superior to competitors such as coal, batteries and even hydrogen by the energy content in relation to weight and size, in addition to being easily transported and stored. Oil still is essential for transportation and the exchange of goods, industrial specialization and productivity gains that have permitted major improvements in real wages and living standards in large parts of the world.[12] Correspondingly, disruptions of oil supplies and abrupt oil price rises have caused economic setbacks, social distress and political discontent. During the 20th century, oil has been a key factor in the outcomes of two world wars. The Allies' access to oil was essential to their victories. The military significance of oil is essentially unchallenged in the 21st century.[13] Nuclear submarines and aircraft carriers do not diminish the need for oil for navies, armies and air forces, and hence, the military significance of oil.

Against this backdrop, security of oil supplies has become a major concern in many of the world's capitals. The actual meaning varies from secure volumes to secure prices. From the perspective of oil consumers, secure oil supplies should mean stable supplies at stable prices, or preferably, rising supplies at stable prices. Measured by turnover, oil remains the world's most important traded commodity. It is an essential input factor in the transportation sector. For consumers, the most costly oil is the volume that is needed, but not available.

Oil revenues are essential to the entire Gulf economy, meaning that both oil price levels and market share represent a potential for conflicts. Historically, tension over oil market shares and prices has marked relations between Iran, Iraq, Kuwait and Saudi Arabia, being one of the factors driving Iraq in 1990 to occupy Iraq. The breakdown of the OPEC consensus in 2011, with Saudi Arabia unilaterally expanding volumes, indicates that such conflicts are still real. This is corroborated by disagreement over the succession of the OPEC Secretary General. Late 2012, Iran and Iraq lined up against Saudi Arabia on the issue of the new OPEC Secretary General.

Cumulative cleavages enhance the conflict potential. Confessional differences between Sunni and Shia in Bahrain as well as in Iraq, Syria and Saudi Arabia also represent social and economic disparity, so that the *de facto* inclusion of Bahrain as a province of Saudi Arabia compromises the political stability of the latter, enhancing the potential for foreign rivals, whether Iran or Iraq, to play the Shia card to weaken the Saudi regime. Both Iran and Iraq are competitors to Saudi Arabia in the oil market and, as mentioned above, their oil price preferences are considerably higher, given larger populations and a less favourable financial position. Hence, Saudi expansion of oil supplies may appear as a threat to the economic stability of Iran and Iraq.

Finally, the republican regimes of Iran and Iraq contrast with the absolute monarchy of Saudi Arabia and posing an ideological challenge. Arab, nominally secular Iraq may prove to be a tougher test for the Saudi regime, especially in case a dispute over economic matters, such as oil, should ignite tension. In addition, Kuwait is subject to political strife over the power of elected bodies versus the royal family.

Political Strategies and Economic Strength

Historically, economic distress and disputes have provided incentives for conflict and wars in the Middle East and in the Gulf region. In the aftermath of the First World War, the carve-up of the Ottoman Empire caused strong tensions between Britain, France and the United States over the control of oil. The outcome was the creation of Iraq, with French and US oil stakes recognized in a British mandate. After the Second World War, oil was a major ingredient in the 1951-1953 Iran crisis and the overthrow of democratic government with the instauration of monarchy at the behest of Britain and the United States. In 1973-1974, following the Yom Kippur War, quadrupling oil prices fundamentally changed economic power relations in the Middle East, a trend strengthened by the 1979-1980 Iranian Revolution. The subsequent 1986 oil price decline unmade much of these gains, except for the few oil exporters with the potential to raise volumes in order to make up for lower prices. This pitted Saudi Arabia, Kuwait and the other Gulf states against the rest, especially Iran and Iraq, which were at war against each other. The conflict between populous Iran and Iraq, on one hand, and the less populous Gulf Stares, on the other hand, over oil prices and volumes had been manifest since the early 1970s. By 1990 it caused war.

By 1990, after the war with Iran had ended, Iraq was approaching bankruptcy, unable to pay for food imports and the military, and service a huge foreign debt by limited oil revenues due to low oil prices. For the Saddam Hussein regime the options were to move out, to step down, or to move forward, occupying Kuwait, with the hope of recovering not only oil, but also financial assets.

The Iraqi occupation of Kuwait triggered the first Gulf War in 1991 and the subsequent no-fly zone over northern Iraq that permitted Kurdish home rule. Since then, oil exports have made up the economic backbone of the Kurdistan autonomy. It is currently under threat from the conflict between the KRG and the federal Iraqi government, with the potential of provoking an armed conflict between Turkey and Iraq. Here, oil is the direct bone of contention, but Turkey also has an interest in natural gas from Iraqi Kurdistan. In this case, the strife over the control of oil coincides with the ethnic cleavage between Iraqi Kurds and Iraqi Arabs, and with rivalry between Turkey and Iraq. The cumulating conflicts carry a risk of military confrontation, also affecting the balance in the Gulf.

The present tensions in the Gulf region, with repercussion to the Middle East and North Africa at large, can be seen from a geopolitical perspective as an outcome of a gradual and painful transition from unipolarity, led by an unchallenged United States, to multipolarity, where US leadership is eroding and is increasingly challenged by a number of other countries, not the least China, India, Russia and Turkey that also seek positions in the Gulf.[14] In this competition for power and influence in the Gulf region, the cards are unevenly distributed. Among the two major contenders, China has much money and has chosen a mercantilist strategy to gain footholds, and the United States, with less money, has so far chosen a militaristic strategy to retain its foothold.

Iran and Saudi Arabia are essential pieces in this struggle. Iran represents the break with the past, unipolar order and resistance to attempts to restore US hegemony in the region. This position draws support from countries, not the least China and Russia that do not wish to see Iran falling back into the US sphere of influence. The Iranian strategy seems to be to withstand economic sanctions in the short run, and in the longer run to gain more political influence in the Middle East, not only in the Gulf region.

Saudi Arabia is the counterpart, a conservative monarchy closely aligned with the United States and attempting to counter Iranian influence. This position draws support from fellow conservative monarchies, essentially Qatar and the United Arab Emirates, although less so from Kuwait. The Saudi strategy seems to be to fight Iran's supporters and to support any force opposing Iran. This is essentially a short-term strategy and the issue is what long-term regional strategy that Saudi Arabia might have, if any. Even if Saudi Arabia currently enjoys a comfortable economic situation, it does not have unlimited financial means facing serious youth unemployment. Since the *coup d'état* in Egypt in 2013, Saudi Arabia has provided billions of dollars in assistance. The generous amount does not appear sufficient to stabilise Egypt and it is questionable to what extent Saudi Arabia would have the means to continue support on such a lavish scale. Any destabilisation of Egypt would also affect Saudi Arabia. The further risk is that instability in Libya would affect Egypt, and indirectly Saudi Arabia.

Qatar, as a small and wealthy country, has been even more active in supporting the Muslim Brotherhood. The long-term strategy remains an open question. The United Arab Emirates, UAE, is in a similar position, although at least one of the members, Dubai, benefits economically from circumventing sanctions against Iran. Both Qatar and the UAE have lavish financial resources enabling them to project influence that is disproportionate with their small populations. The issue is what long-term strategic objectives they might pursue, but in any case they differ. Currently, Qatar is at odds with Saudi Arabia and the UAE for both ideological and business reasons. As a supporter of the Muslim Brotherhood, Qatar pursues a policy antagonistic to Saudi Arabia. In business, Qatar has huge reserves of natural gas that Saudi Arabia needs, but two have not managed to agree on a contract.

Kuwait has taken a more neutral stance, due to a history of enjoying Iran's support against Iraq and because of a large domestic Shia population. Kuwait's strategy, in the short run as in the long run, seems to be to stay out of conflicts and not to compromise relations with neither Iran nor Saudi Arabia. The issue is how Kuwait might choose in case of more direct conflict between Iran and Saudi Arabia.

The Bahrain crisis is part of a more comprehensive Gulf crisis. It has been presented as a conflict between Sunni and Shia, but it is also a crisis of legitimacy of an autocratic government with regional and geopolitical repercussions. From a geopolitical perspective it appears as an attempt by the United States acting through Saudi Arabia to contain Iran, also with a parallel to Syria. Even if the Bahrain crisis has been presented as a clash between Iranian and Saudi influences, it is also a confrontation between different Arab allegiances, Sunni to Saudi Arabia, Shia to Iraq. Saudi Arabia is actively supporting the Sunni minority in Iraq, which in return might see an interest in supporting the Shia population of the Gulf states, not only in Bahrain, but possibly also the Shia in Saudi Arabia's Eastern Province. From this perspective there is a risk that Iran and Iraq together will confront the conservative Gulf monarchies, with Saudi Arabia as the major adversary backed by the United States.

Even if the uprising in Bahrain has developed because of domestic grievances, Saudi Arabia fears that majority rule in Bahrain would open for Iranian influence, also among its own Shia minority. The issue is what might loosen the stalemate in Bahrain. In the summer of 2014, political reform has been proposed, but its fate is uncertain.

In Syria, the situation is more complex, with a minority rule backed by Iran fighting a diverse opposition coalition backed by the west as well as Saudi Arabia and Qatar. So far, Syria's minority government has been an Iranian ally, which incites the conservative Sunni monarchies to support regime change. Even if the Assad regime has proven more resilient than expected at the outbreak of the civil war, until recently the prospect was that a Sunni based majority regime would take over, the question being when and how.

For Iran, a regime change in Syria would imply a loss of an ally. In that case, Iraq a Shia majority and potentially rich neighbour would become even more important as a partner for Iran. Any loss of Kurdistan would strengthen the Shia character of Iraq. This eventuality might, however, incite Saudi and Qatari efforts to bolster the Sunni opposition in Iraq in order to pre-empt greater Iranian influence. The risk is yet another arena of armed conflict by proxies between Iran and Saudi Arabia, possibly civil war and a partition of Iraq that would leave the remainder at the mercy of Iran. The issue is what strategies Saudi Arabia and Qatar conceivably could pursue over Iraq. The issue is also what political objectives Iran pursues in relation to Iraq in addition to strengthening bilateral economic ties.

Iraq – Fractures in the Cornerstone

Iraq is the cornerstone of the Middle Eastern state system, bordering on six neighbours. Consequently, the integrity of Iraq directly affects Iran, Kuwait, Saudi Arabia, Jordan, Syria and Turkey in their foreign relations as well as their domestic politics. Historically, the strong, centralized Iraqi state was experienced as a problem, if not a threat by each of the neighbours. Today, any partition or further weakening of Iraq would have profound repercussions in the rest of the Arab world and the Middle East at large.

Indeed, in the aftermath of the First World War, Iraq was established as a composite state, joining three distinctive regions, as mentioned to suit British and French oil interests. Mixing Shia and Sunni citizens in central Iraq, especially in Baghdad, took decades of migration from the countryside to the city. Since the beginning, Iraq's three components have been kept together by force, most of the time by minority Sunni governments. The counterpart to internal repression was external aggression. Until 1990 Iraq was the leading military power in the Middle East. In oil matters, Iraq was a notoriously unreliable member of OPEC that would steadily disregard production quotas. Against this backdrop, Iraq under Saddam was seen as a threat by the Gulf neighbours, and rightly so.

Iraq could be a rich country, provided it gets its politics right. It has potential for agricultural development and for feeding its population, besides that for industrial development and large oil resources. Since independence, it has developed an urban merchant class. The long-range objective of Iraqi economic policy in the 1960s and 1970s was to develop industry, to diversify and reduce dependence on oil.[15] The objective was not reached, essentially because of the first oil price rise that raised the share of oil in its overall economy. For several years Iraq did, however, keep its oil production below capacity and output did not expand during the years immediately following the first oil price rise. Iraq also, until Saddam Hussein took power, tried to keep oil production and its revenues at a level that the country could absorb, preferably in investment projects with a reasonable return.[16] In the late 1970s Iraq embarked on a capital-intensive economic strategy outside oil, neglecting agriculture and food supplies.[17]

After Saddam Hussein took power in 1978 the country raised military expenditure. The military effort increased in 1980 after Iraq attacked Iran, and when higher oil prices were being earned. During the 1980s Iraq increased its oil production substantially, to finance the war against Iran and keep civilian life running with as little disturbance as possible. In the 1990s, sanctions and isolation made the regime liberalize the economy, selling off state industries, services and agricultural land.[18] President Hussein's immediate power base was a small number of family, clan and regional affiliates, but it also comprised a broader community of privileged persons. Their loyalty was based on government positions and newly acquired private property.

Before the Iran War, Iraq had achieved a comparably high level of economic and social development.[19] Even if the income base was almost entirely dependent on oil, Iraq ranked internationally among upper middle countries such as Greece and Venezuela. The relative prosperity was manifest through good nutrition standards, with a per capita daily intake of about 3000 calories. The education system functioned well and illiteracy had been almost obliterated.

With the world's second largest proven oil reserves and the potential for successful exploration, Iraq enjoys a unique freedom of choice in oil policy.[20] Whatever option Iraq chooses, it will affect the world oil market and oil prices. After normalization, Iraq will attract the world's leading oil companies as investors. Provided political conditions are more palatable, Iraqi expertise is likely to return, strengthening the competence base of the national oil industry, with oil production expanding quickly, essentially financed by foreign investment. The issue is rather in what way quickly expanding Iraqi oil exports will influence the market and, essentially, what the eventual response will be from the other major Middle East oil exporters. With a huge resource base and low costs, Iraq is one of the few OPEC countries with the potential ability for increasing oil revenues by expanding volumes to offset declining oil prices. Iraq's preference may be to enjoy higher oil revenues through a combination of higher volumes and comparably high oil prices, eventually at the expense of other major Middle East oil exporters. Geopolitics, essentially the US presence in the Gulf and the military balance, will decide the outcome. This observation stresses the potential for oil market instability associated with Iraq.

Historically, Iraq has been a recurrent defector from OPEC agreements. This was the case in the 1970s and 1980s until the Gulf conflict. It is likely to be the case again, once sanctions are lifted. Iraq has the resources and the incentives not to obey OPEC agreements, unless the country gets into a dominant position where it can challenge that of Saudi Arabia in influence as well as in volume. For the rest of OPEC, Iraq represents a risk: any generous terms it might grant the international oil companies would strengthen their bargaining position in relation to the other oil-exporting countries and their national oil companies. After years of sanctions, wars and occupation, Iraq has reasons to aim for a free ridership in OPEC, going for a higher market share and leaving price stabilization to its neighbours; alternatively it may decide to test the competitive edge. Ultimately, this is a recipe for conflictual relations with its Middle East neighbours and competitors.

In hindsight, the beginning of the end for Iraq was marked by the decision to go to war against Iran in 1980. Iraq had territorial grievances against Iran concerning the waterways of Shatt al-Arab. The Sunni minority Saddam regime saw the new Islamist regime in Teheran as a serious threat to its control of its own Shia majority population. It therefore had reasons to weaken or even overthrow the Islamist republic. Moreover, the US Carter government tacitly encouraged Iraq to attack Iran, hoping that the Islamist regime would collapse and the hostages be released.[21] The Arab Gulf countries actively instigated, funded and encouraged Iraq's war effort against Iran.

The eight-year war turned out to be a catastrophe for Iraq, not the least as Israel chose to discretely supply Iran with arms in order to forestall a victory that would have made Iraq the leading regional power.[22] After the war, Iraq was ruined, with a huge debt burden that it could not service. In the meantime, oil prices had fallen and both Iraq and Kuwait were producing above OPEC quotas. A large part of the debt had been contracted to the rich Gulf States, Kuwait, Saudi Arabia and the United Arab Emirates, which refused to renegotiate the Iraqi debt. Consequently, the Iraqi regime found itself in inextricable financial difficulties.[23] The mirror image was across the border. Kuwait appeared as the world's largest treasury with huge oil reserves as well as ample financial reserves.[24]

The Iraqi regime in 1989–90 suffered from falling real oil revenues. With a commitment to high military expenditure, it was unable to both import food and service the foreign debt. Even if the Iraqi debt to other Arab countries was unenforceable, the Iraqi government was in a precarious financial situation. This was one of the motives for choosing external aggression, attacking Kuwait, to eventually avoid a social and political upheaval with the prospects of a bloody end to the regime. A third way out, economic and political reform, at the time seemed out of the question. The outcome might have been feasible, triggered by a military coup to replace Saddam with a leader more acceptable to the richer neighbours and creditors and who would have withdrawn voluntarily from Kuwait.

Against this backdrop, the Saddam regime in Iraq had the choice between defaulting and eventually vacating power, with unpredictable internal and external consequences, or making a leap forward, grabbing the oil and money of Kuwait. Controlling Kuwait would have strengthened Iraq's position as the leading regional power, threatening Saudi Arabia, with a *de facto* direct and indirect control over a considerable part of the world's oil supplies. This eventuality was unacceptable to the neighbouring Gulf States as to the world's oil importers. The following brief Gulf war early 1991 marked the beginning of Kurdish autonomy in northern Iraq and the entry of the United States as the leading military power in the Middle East.

Since 1991, Iraq has been steadily weakened through sanctions, the 2003 war and the ensuing US occupation. In 2011, the United States left a country that is profoundly divided with political institutions that are barely functional, harmed by rampant corruption, strife and serious security problems. Ten years after the US invasion, electricity supplies are deficient. Oil investment is set back by government incompetence and deficient security.[25] As mentioned, the risk of a partition appears imminent. The risk is underlined by the presence of US oil companies in the autonomous Kurdish region contrasting with Chinese and Russian oil companies operating in southern Iraq.

The US occupation shifted political power from the Sunni minority to the Shia majority. Iraq today has two minorities, Kurdish and Sunni, whose relations are strained. The Kurdish population suffered heavily from Saddam regime repression, with which they associate the Sunni Arabs. The Kurdish Regional Government is claiming largely or partly Kurdish areas in northern Iraq, especially Kirkuk, in conflict with Sunni Arab claims.

Even if the United States pays lip service to the principle of a unitary Iraq, corporate and Israeli interests push in the opposite direction. Even if the US government would prefer not to be seen as promoting Kurdish separatism and the ensuing partition of Iraq, it is unlikely to oppose such a development. Indeed, in the years following the occupation, strong voices in the United States did call for splitting up Iraq into three parts in a loose federation or even with outright partition.[26] Moreover, during the US occupation, large-scale ethnic cleansing took place among Iraqi Arabs, leading to a more distinct separation of Shia and Sunni populations. Thus, the Arab central part of Iraq has shifted from a region consisting of confessionally mixed areas to a region of more homogeneous areas that could more easily be split up. This is an important long-term effect of the US occupation after it has ended, with potentially critical consequences for Iraq as a state.

Benefits for Iran in Iraq

Beyond any doubt, Iran has benefited the most from the US-led Iraq war and the ouster of the Saddam Hussein regime, another show of the law of unintended consequences. Evidence is in booming trade between the two countries, Iranian support of Shia welfare organizations and militias in Iraq, as well as in close personal links with Iraqi politicians who spent years in exile in Iran. Against this backdrop, Iran is a force in Iraqi politics.

Iraq has without much attention become Iran's major partner in the Middle East, possibly more valuable even than Syria. Iraq does not observe sanctions on Iran. Allegedly, volumes of Iranian oil find their way to the market through Iraq. In the short term the Maliki government appears well entrenched in spite of political unrest and occasional terrorism, to which it responds by increasingly authoritarian means that cause public concern. High oil revenues contribute to stability as well as to rampant corruption that incites public anger.

Within the framework of formally democratic institutions, the multiple cleavages and conflicts of Iraqi society play out openly. The issue is not only Shia versus Sunni or Arab versus Kurd, or secular versus religious, but also Shia versus Shia, Sunni versus Sunni and Kurd versus Kurd, as well as secular versus secular. Within this complex setting, coalitions and alliances emerge and dissolve. In addition, strong personal rivalries season Iraqi politics. In 2011, Prime Minister Maliki issued an arrest order on his deputy and his vice president. Both are Sunni, but belong to a political grouping led by secular Shia leader Iyad Allawi, a former prime minister. In 2012 Shia leader Moqtada al-Sadr visited Iraqi Kurdistan for protracted talks with Sunni Kurdish leader Massoud Barzani in the presence of the secular as well as religious Sunni opposition.

In Iraq the forces that made the Arab Spring can more readily play out in a liberal institutional context, although grievances are serious. An authoritarian twist by the Maliki government would risk provoking public disorder. Moreover, Saudi Arabia and Turkey have an interest in strengthening Iraq's Arab Sunni politicians, although they represent a minority. This is a risk for Iraq's political stability, possibly also its integrity.

Iraq apparently aims at once more to become an important military power in the region. The Maliki government acts to strengthen Iraq's integrity and to pre-empt Kurdish endeavours at greater autonomy within a larger territory. So far, there is no agreement on new conditions for oil exports from Iraqi Kurdistan. In the autumn of 2012 Baghdad placed a military command in Kirkuk, requesting that the Kurdish Peshmerga stay away. The threat to its integrity induces Iraq to seek closer ties with both Iran and Syria. Iraq's interest is also to maintain good relations with Syria in order to prevent closer ties between Syria and Iraqi Kurdistan. As already pointed out, the integrity of Iraq is critical to the state system of the Middle East. Any implosion or disintegration of the Iraqi state risks provoking protracted disorder and strife involving all neighbouring countries, also the Gulf, compromising oil supplies.

Even if Iran and Iraq have different and to some extent competing interests, they also have salient common interests in relation to the oil market, Saudi Arabia and Syria. As for oil, Iran and Iraq have a common interest in high oil prices and in high market shares for themselves, if necessary at the expense of Saudi Arabia and the smaller Gulf states. As for Syria, Iran and Iraq have a common interest in keeping the current regime and in averting either a Sunni regime dominated by the Muslim Brotherhood or a pro-Western regime.

Therefore, Iran and Iraq have a common ground for strained relations with Saudi Arabia, Turkey and the United States.

The US occupation of northern Iraq guaranteeing home rule for the Kurdish region has reopened the Kurdish issue. Oil resources would provide an economic basis for a Kurdish state. As pointed out, complete separation of Kurdistan from Iraq can no longer be ruled out, in which case the political map of the Middle East would see a major revision. Control of Kirkuk and its oil would be a critical issue. Repercussions would be profound for oil politics and markets, as for the balance of power in the Gulf.

The United States – the Historical Hegemon Challenged

As the historical hegemon, the United States remains the leading military power of the region. Even if US oil imports are declining, the United States retains strong economic interests in the Gulf. It is in the US interest that Saudi Arabia and the other Gulf States continue pricing oil in US dollars, although their major market is China. Likewise, the United States has an interest in Saudi Arabia and the other Gulf States to buy US military equipment and services, and that they continue investing their financial surpluses in US securities. The issue is what interest Saudi Arabia and the other Gulf States would have in continuing these policies.

Until the late 1970s, the USA had close relations with the governments of all regional big powers—Egypt, Iran, Saudi Arabia and Turkey—and used them as conduits for projection of influence in the region. The modus operandi was classical bridgehead imperialism: when interests diverged, local governments often sided with the hegemon and not their own domestic constituencies. By 2013, three of them are conducting foreign policies more or less independently of the USA, and the fourth—Saudi Arabia—is critical to Washington's evolving regional policy. The strong rise in Saudi military expenditure seems to signal that the Kingdom is responsible for its own security and no longer depend on US guarantees.[27]

China is the challenger and has become the major trade partner for the Gulf region. China is the major outlet for Saudi and other Gulf crude oil, the major investor in Iraq and Iran, and Iran's largest trade partner. To assert influence, China essentially uses commercial and economic means as part of a mercantilist strategy to win access to raw materials and markets, whereas the USA essentially uses armed forces as part of a military strategy. China's interest is to conduct trade in yuan, not in US dollars, which is already the case for trade Iran. As the major buyer of Saudi oil, it may be matter of time before China requests the ability to pay for Saudi oil in yuan. The issue is what attraction that might have for Saudi Arabia. A final question is to what extent China's economic clout in the Gulf might be employed to prevent conflict and promote regional stability, which is essential for the world's major oil importer.

China's oil imports grow as US oil imports recede. This may reduce tension and the conflict potential over access to oil. As the centre of import gravity in the oil market shifts from the North Atlantic to South and East Asia, important economic and political consequences follow. The flows of manufactured goods from China to compensate for the oil flows are shifting the patterns of world trade in general, not just oil and other raw materials. The flows of investment and payment are altering the patterns of international finance, likely to be reinforced by the gradual internationalisation and convertibility of the Chinese currency. Finally, the flows of oil, manufactured goods and money are creating new patterns of political allegiances. China is already the leading economic power in the Middle East, with growing political influence. Any drive by the United States and the European Union to impose harsher sanctions on Iran would serve China's interests well.

Within this setting, the United States appears to be heading for increasing marginalisation in world oil trade. With oil imports receding, the US market will be less important to the major oil exporters, but the US trade deficit is also diminishing. The issue is to what extent higher self-sufficiency in oil and budgetary constraints will reduce the US military commitment in the Middle East, or by contrast, lead to an even more unilateral pro-Israeli stance, further unsettling the Middle East and the oil market.[28] Anyway, the signal from Washington is that with greater self-sufficiency in oil, the United States will be able to give a higher priority to diplomacy over military means in pursuing its interests in the Middle East.[29]

A major handicap for US Middle East policy, as with US foreign policy in general, is shallow knowledge and lack of insight into the external world among politicians and even high-ranking civil servants. This problem is particularly pertinent in relation to Middle East policy, which is strongly influenced by pro-Israeli interests. Pro-Israeli interests are particularly influential in the House Foreign Relations Committee, where the essentials of US Middle East policy are carved out.

Insight into international oil matters is superficial, as witnessed by former US Defence Secretary and CIA Director John M. Deutch who in 2011 erroneously wrote that in 1973 the twelve members of OPEC announced an oil embargo on the United States. Indeed, the embargo was called by the Organization of Arab Petroleum Exporting Countries, OAPEC, not OPEC. The six non-Arab members of OPEC, i.e. Ecuador, Gabon, Indonesia, Iran, Nigeria and Venezuela, did not take part in the embargo, nor did Iraq. Indeed, the majority of OPEC member states did not contribute in an action called for by another organization. The statement referred to reveals critical lack of knowledge of the outside world, also on matters pertinent to the United States, at the very core of political decision-making in Washington. The neoconservative ideologues who initiated the Iraq War did not have any profound insight into Middle East affairs, which is one reason why their adventure was doomed from the beginning. Unfortunately, there is in Washington a certain continuity of forcefulness mixed with ignorance. Iran potentially could be the next victim.

Indeed, in relation to Iran, US policy makers seem to be the victims of their own propaganda depicting the ruling politicians in Teheran as irrational "madmen". Successive

US administrations since the 1979 Islamic Revolution have shown limited effort to analyse and understand the rationality, the objectives and the methods of the Iranian regime.

By uncritically supporting US policies on the Middle East, the EU has chosen irrelevance in world affairs, neglecting distinctive European economic and political interests. At an earlier stage, the euro appeared as a rival to the dollar, also in oil trade; currently the Chinese yuan has that position.[30]

The United States and China have a common interest in Gulf stability, an interest shared with Europe, India and many other countries. Currently, and probably for many years, the United States has an interest in a comparatively high floor for oil prices to safeguard its domestic high-cost shale oil industry, while China as a large and growing importer has an interest in moderate oil prices. Against this backdrop, a kind of a US-Chinese condominium in the Gulf might emerge, with the United States providing military patrolling and China the financial means of discipline and control. Saudi Arabia already has industrial links with both, as Saudi Aramco has invested in refineries in China as well as in the United States in order to secure outlets. A possible outcome might even be for Saudi Arabia to invoice oil in yuan to China, in dollars to the United States and perhaps the rest of the world, or in euros to Europe.

Finally, the natural gas deal and oil agreement concluded between China and Russia in May of 2014 could also have repercussions for Gulf oil trade. Russian energy exports to China are increasing fast. The likelihood is that the trade will be conducted in yuan and roubles, not in US dollars. That would challenge the position of the US dollar as the world's reserve currency as well as the financial leeway and power of the United States.[31] With a competing petrocurrency, the attractiveness of the US dollar to central banks is likely to diminish.

Notes

1 Chulov, Martin, "Saudi Arabia moves to settle differences with Iran", *The Guardian*, 13 May, 2014.

2 Parsi, Trita, "No, Sanctions Didn't Force Iran to Make a Deal", *Foreign Policy*, 14 May, 2014.

3 Brian McBeth, *British Oil Policy 1919–1939*, London 1985, Frank Cass, 1985, pp. 24ff.

4 Brian McBeth, *British Oil Policy 1919–1939.*

5 David Fromkin, *A Peace to End All Peace* (New York: Avon Books, 1989), p. 375.

6 McBeth: *British Oil Policy 1919–1939*, p. 32.

7 Ofra Bengio, 'Iraq's Shi'a and Kurdish communities: from resentment to revolt', in Amatzia Baran and Barry Rubin (eds), *Iraq's Road to War* (New York: St Martin's Press, 1993), pp. 51–66.

8 Todd, Emmanuel, *Allah n'y est pour rien*, Paris 2011, arretsurimages.

9 Hossein Askari, "Dark outlook for post-sanctions Iran", *Asia Times Online*, 14 May, 2014, www.atimes.com.

10 Shahriar, Amin, "Tehran treads social liberties tightrope," *Asia Times Online*, 19 May 2014, www.atimes.com.

11 Richard A. Muller, *Physics for Future Presidents*, New York 2008, Norton, p. 15.

12 Matt Ridley, *The Rational Optimist*, New York 2010: Harper Perennial, p. 56.

13 Le Bideau, Hervé, *Le pétrole, enjeu stratégique des guerres modernes*, Paris 2010: Editions L'esprit du livre.

14 Brzezinski, Zbigniew, *The Grand Chessboard*, New York 1997, Basic Books, p. 209.
15 Phebe Marr, *The Modern History of Iraq*, Boulder, CO: Westview Press, 1985, pp. 247ff.
16 Fadhil Al-Chalabi and Adnan Al-Janabi, 'Optimum production and pricing policies', *Journal of Energy and Development*, Spring 1979, pp. 229–58.
17 Amuzegar: *Managing the Oil Wealth*, p. 133.
18 Aziz-Chaudhry: 'Economic liberalisation and the lineages of the rentier state', p. 368.
19 Anthony H. Cordesman and Ahmed S. Hashim, *Iraq*, Boulder, CO: Westview Press, 1997, p. 127.
20 Derek Brower, "Will Iraq be the next oil superpower?", Petroleum Economist, December 2012/January 2013, pp. 14-19.
21 Dilip Hiro, *The Longest War*, New York 1991, p. 72.
22 Trita Parsi, *Treacherous Alliance*, New Haven 2008, Yale University Press, p. 119.
23 Véronique Maurus, « Bagdad est aux prises avec des difficultés financières insurmontables », *Le Monde*, August 3, 1990.
24 Véronique Maurus, « Le Koweit La plus grande banque mondiale » ,*Le Monde*, August 4, 1990.
25 "MENA Security Deterioration Sounds Oil Investment Alarm", *Middle East Economic Survey*, May 3, 2013.
26 "US weighs price of Iraqi territorial integrity", *Financial Times*, November 6, 2006.
27 "What message is Saudi Arabia sending with war games?", *Al-Monitor*, 20 May 2014, www.almonitor.com.
28 Parsi, Trita, *Treacherous Alliance*, London 2007: Yale University Press.
29 *Oil Security 2025*, Washington, D.C, 2014, The P.X. Kelley Center for Energy Security.
30 Subramanian, Arvind, *Eclipse*, Washington, D.C. 2011: Peterson Institute for International Economics.
31 Halligan, Liam, "Russia-China gas deal could ignite a shift in global trading", *The Telegraph*, 25 May 2014.

4

Prospects for Nuclear Power in the Gulf Cooperation Council (GCC)

Ali Ahmad and M.V. Ramana

Introduction

In a 2006 summit meeting in Riyadh, the leaders of the six Gulf Cooperation Council's (GCC) states collectively declared their intention of starting a joint nuclear energy development program.[1] Since then, Bahrain, Kuwait, Oman and Qatar appear to have walked away from their nuclear plans in the wake of the multiple accidents at Japan's Fukushima nuclear plant. The UAE and Saudi Arabia, on the other hand, have continued with their nuclear plans, albeit at different paces. The first two units of the UAE's first nuclear power plant (Barakah) are currently under construction and, as of May 2014, are scheduled to be connected to the grid in November 2016 and 2017 respectively.[2] Saudi Arabia has continued to reiterate its commitment to the acquisition of nuclear power and in May 2012, the King Abdullah City for Atomic and Renewable Energy (KA-CARE) projected that by 2032, out of a total 123 GWe of electricity generation capacity in the country, 18 GWe would be contributed by nuclear power.[3]

Across the GCC, officials offer multiple motivations to construct nuclear reactors. Perhaps the most commonly articulated one is a desire to get away from the near-complete reliance on hydrocarbon resources to produce electricity and desalinate water. Such reliance is seen to lead to the depletion of national oil and gas reserves but is also perceived to have an opportunity cost associated with forgone export revenues. Others suggest that interest in nuclear power is also a response to Iran's acquisition of nuclear technology. It has been argued, for example, that, "GCC states want to show Iran, their own people, and the broader world that Arabs also have the prowess and power attributed to nuclear technology".[4] Whatever the motivations behind nuclear power might be, the acquisition of nuclear technology will entail major economic and non-economic costs.

This paper studies the potential role of nuclear power for the GCC countries, and compares the economics of generating electricity from nuclear reactors to alternate sources such as natural gas and solar energy. This analysis also considers the opportunity cost associated with using natural gas in comparison to nuclear power. While many of the GCC states are not pursuing nuclear power currently, our analysis would be applicable to them if they do so in the future.

Energy Statistics in the GCC Countries

The average behavior with regards to energy consumption and its relationship to economic activities and efficiency levels in the GCC states is shown in Fig.1. A parallel coordinates chart shows five efficiency-related indicators—electricity consumption per unit of gross domestic product (GDP); electricity consumption per capita; carbon emissions per unit of GDP; carbon emissions per capita, and the percentage of electricity consumed in the residential sector in the GCC—for the GCC states and the global average based on 2012 data from the World Bank and the International Energy Agency.[5] On a per GDP basis, the GCC states perform reasonably well on their energy and emissions profiles compared to world's average because of their high GDP. On a per capita basis, however, the GCC states are among the highest electricity consumers in the world. Kuwait and Qatar are currently ranked forth and eighth largest electricity consumers per capita in the world respectively. Similarly, the per capita carbon emissions in the GCC are among the highest in the world with Qatar ranked first and Kuwait forth. Electricity consumption in the residential sector reflects both the impact of growing population and urbanization in the GCC but also the potential for energy efficiency improvements.

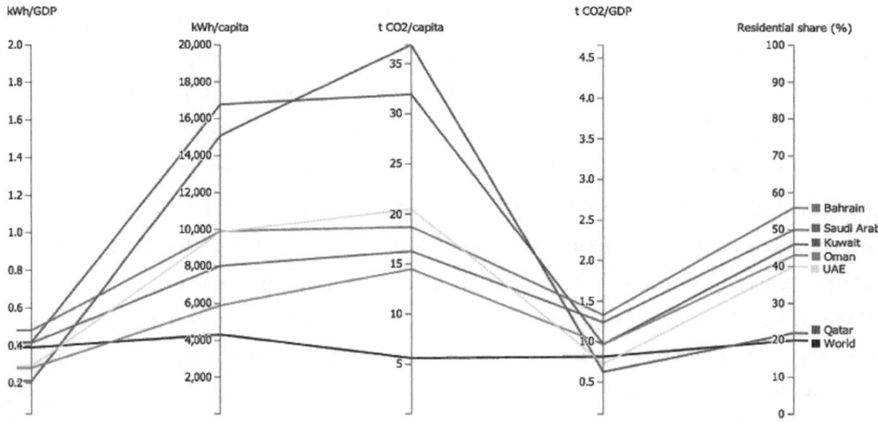

Figure 1 **Indicators of Energy Efficiency in the GCC**

Figure 2 shows five indicators related to electricity generation in the GCC states: installed capacity; electricity consumption average growth rate; electricity from fossil fuel; electricity from natural gas; and gas exports. These figures are based on 2012 data from the U.S. Department of Energy's Energy Information Administration and the International Energy Agency. In terms of generation capacity, which correlates with electricity grid size, the GCC

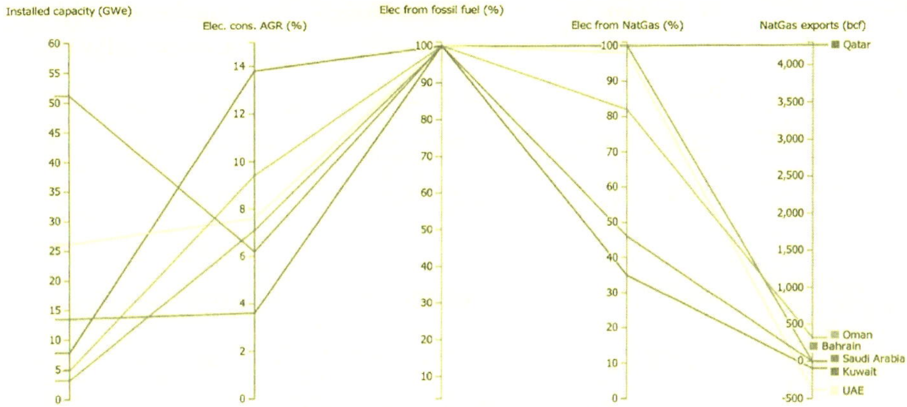

Figure 2 **Electricity Indicators in the GCC**

states can be divided into two groups: Saudi Arabia and United Arab Emirates (UAE) with relatively large grids; and Kuwait, Oman, Qatar, and Bahrain with relatively smaller electricity grids. This distinction between large and small grid capacities has a bearing on the compatibility of large nuclear reactors. The average growth in electricity consumption is generally high apart from Kuwait, which has been growing at below 4% per year. Qatar has the highest average rate of increase in electricity consumption, just below 14%. This is due to recent expansions in infrastructure and a sharp increase in population since 2005.

For electricity production, the GCC relies solely on oil and gas. Bahrain, Qatar and the UAE produce all of their electricity from gas-fired power plants while Oman, Saudi Arabia and Kuwait currently produce 82%, 46% and 35% respectively of their electricity using natural gas. Currently the UAE and Kuwait are the only countries in the GCC that import natural gas. Saudi Arabia and Bahrain completely utilize their natural gas domestically and do not export any.

Nuclear Technologies

There are multiple reactor technologies at different stages of readiness available for the export market. Many of the reactors that have already been sold and are being operated or under construction have relatively large power capacities (Table 1).[6] More detailed technical descriptions of reactor technologies and their relevance to the Middle East market can be found in the literature.[7]

In the GCC itself, the South Korean APR-1400 technology is the only one currently under construction. The UAE's decision to purchase the AP-1400 reactor from South Korea, in preference to other vendors from France, Japan and the United States, did come as a surprise to many. By securing the contract, South Korea moved towards being perceived as a credible nuclear vendor. Its success in the case of the UAE is said to have to do, in part, to

Country	Technology	Reactor Design	Approximate Capacity (MW)
South Korea	PWR	APR-1400	1400
France	PWR	EPR	1600-1700
USA/Japan	PWR	AP1000	1000
Russia	PWR	VVER	1000
USA/Japan	BWR	ABWR	1350
India/Canada	PHWR	PHWR/ACR-700	700

***Table 1* Current Reactor Designs**

being flexible and meeting the UAE's bid requirements;[8] attractive financing and offering to conduct training programs and personnel development of the host country may also have played a part. However, so far, it has not managed to replicate its success in the UAE in other countries in the region or elsewhere.

Saudi Arabia's interest in nuclear power has attracted a number of nuclear vendors. In September 2013 both GE Hitachi Nuclear Energy and Toshiba/ Westinghouse signed contracts with Exelon Nuclear Partners (ENP), a division of Exelon Generation, to pursue reactor construction deals with KA-CARE.[9] The reactor designs proposed include the ABWR, the ESBWR, and the AP1000. The French companies Areva and EDF have also been aggressively moving into Saudi Arabia, and have reportedly started training Saudi engineers and have set up an office in Riyadh to lay the groundwork for a possible bid for tender.[10]

Small Modular Reactors

In recent years, a number of designs called Small Modular Reactors (SMRs), with power outputs of less than 300 MWe, are being promoted by nuclear establishments in many countries.[11] The term Small is used to indicate that the power level is much lower than the average power delivered by currently operating reactors (Table 1). Modular means that the reactor is assembled from factory-fabricated parts or "modules". Each module represents a portion of the finished plant built in a factory and shipped to the reactor site. Modularity is also used to indicate the idea that rather than constructing one large reactor, the equivalent power output will be generated using multiple smaller reactors that allow for greater tailoring of generation capacity to demand.

Although the term SMR might suggest a single reactor type, there are in fact a very wide number of nuclear reactor designs with distinct characteristics being developed. These designs vary by power output, physical size, fuel type, enrichment level, refueling frequency, site location, and spent fuel characteristics. They are also in different levels of development, with some that are in the process of being constructed (e.g. the Russian KLT-40 floating plant) and others that still face major technical challenges that are unlikely to be overcome over the next decade or longer.

A number of nuclear vendors have argued that there are multiple motivations to pursue smaller designs, directed both at large industrialized countries and developing countries. One motivation is the high upfront capital cost of standard reactors, which is beyond the financing capacities of many utilities and countries. Another is to expand nuclear power to countries with relatively small electrical-grid capacities; a gigawatt-scale reactor could destabilize a small grid. Other factors that have been offered as motivations for SMRs are claims to potentially greater safety due to the reliance on passive features, and resulting enhancement in public acceptability. Some SMR designs also promise greater proliferation resistance through one or more features, including reactors with lifetime cores or very infrequent refueling, and the use of fuel forms such as TRISO (uranium fuel coated with multiple layers of different materials) that are harder to reprocess.[12] Specific SMR designs often address one or more of the motivations mentioned above, but it is not always possible to address all desirable criteria simultaneously.[13]

While there is interest in deploying SMRs in several industrialized countries, a major target market for SMRs is to be the developing world, including many countries that have no nuclear capacity at present. The International Atomic Energy Agency (IAEA) suggests that "SMRs can provide an attractive and affordable nuclear-power option for many developing countries" because of their smaller electrical grids, limited infrastructure and restricted investment capability.[14]

These potential benefits of SMRs do not come for free. There is a reason why the current nuclear reactors are much larger than the early prototypes were: smaller nuclear reactors are typically more expensive on a per unit cost basis. Thus, SMRs lose the economies of scale that standard sized (roughly 1000 MW) reactors benefit from. The general rule of thumb governing capital costs of production facilities is known as the 0.6 power rule.[15] That is, the capital costs of two plants of size S_1 and S_2 are related as:

$$\frac{K_1}{K_2} = \left(\frac{S_1}{S_2}\right)^{0,6}$$

Current overnight construction costs for standard sized reactors are estimated to be in the range of $3,000 to $7,000 per kW of capacity. The scaling rule implies that, all else being equal, a SMR with a power capacity of 200 MW would be expected to have a construction cost in the range of $5,700 to $13,000 per kW of capacity. SMR proponents claim that because new reactor designs are different, the comparison with traditional reactor costs is invalid and the scaling law does not hold. They also claim that even if there are diseconomies of scale, these can be compensated by the economic advantages accruing from modular and factory construction, learning from replication, and co-siting of multiple reactors.[16]

Despite these claims, detailed and carefully conducted interviews showed that even experts drawn from, or closely associated with, the nuclear industry expect these reactors to cost more per kW of capacity than currently operating reactors.[17] Based on the average of expert estimates of the extra per kW cost for SMRs, the percentage increase expected ranged from 12% for 225 MWe reactors to a whopping 121% for 45 MWe reactors.[18] These increases are not as significant as the 0.6 power rule would have one expect, but nevertheless any higher cost per kW of capacity translates into a higher cost per kWh of electrical energy. There are also higher fueling costs associated with SMRs.[19]

The suitability of SMRs for the GCC countries will depend on their capacity to host small reactors as opposed to large reactors. The two key indicators are the size of the economy (GDP) and the electricity grid's size. In principle, other criteria including the nature of the demand (peak versus average) and regulatory capacity should also be relevant, but we do not examine those here. All else being equal, if a country has decided to acquire nuclear generation capacity, then as long as its electrical grid has sufficient capacity to host a large reactor, one might expect that from an economic standpoint, the country should prefer to purchase one of the designs currently being constructed (Table 1) rather than an SMR because of the scale disincentives discussed above.

However, if the grid size is less than about 10 GW, then a large reactor is not recommended. There is one caveat: this criterion assumes implicitly that each country is on its own. But two or more countries could integrate their grids and thereby increase the total electrical capacity that flows through the grid. If this is done, then generating facilities in the other country or countries can help provide electricity when the nuclear reactor is shut down for maintenance or refueling.

It may also be the case that the grid is adequately large, but the state under question, or any private utility within the country, may not have the requisite economic capacity to purchase. This is what we measure through the GDP and GDP per capita. Historically, no country starting a nuclear power program has had a GDP below $13 billion (2000 U.S. Dollars) at the time of construction of their first reactor.[20] Likewise, the GDP/capita has never been below $700/capita (in PPP terms). None of the GCC countries fall below these limits and therefore it is unlikely that any of them will have to avoid purchasing a standard sized reactor simply because its economy is too small.

In the case of the GCC countries, therefore, the crucial variable is the electricity grid size. On this basis, if they were to go nuclear, Saudi Arabia and UAE can purchase large reactors whereas Bahrain, Kuwait and Oman and Qatar are better suited for SMRs. And, sure enough, both Saudi Arabia and the UAE are considering large nuclear reactors.

The question that follows is what SMR technologies available to a country that decides to build an SMR? There are currently dozens of SMR designs under development. Some of these are still in the conceptual design phase, many are still in the R&D phase and only four have been licensed or currently are under construction.[21] The details of these SMRs are summarized in Table 2. Thus far, there is little evidence that any of the four remaining GCC countries are considering any of these reactors.

	SMART	KLT-40S	HTR-PM	CAREM
Country of Origin	South Korea	Russia	China	Argentina
Reactor Technology	PWR	PWR	HTR	PWR
Status	Licensed	Under Construction	Under Construction	Under Construction
Electrical Power (MWe)	90-100	35	2X105	27

Table 2 **Small Modular Reactors Currently Under Construction or Licensed**

Economic Challenges of Nuclear Power

Even if a country possesses the two elements discussed earlier—grid size and GDP—energy choices do have to be economically competitive. Other factors do play a part, but economics is usually held out as a key variable to explain energy decisions. And economics has been an important barrier to the construction of new reactors around the world. As the influential interdisciplinary study conducted at the Massachusetts Institute of Technology (MIT) more than a decade ago stated, "Today, nuclear power is not an economically competitive choice".[22] In large part, this was a result of the high capital cost associated with constructing a nuclear reactor. The problem posed by high construction costs is compounded by uncertainty. Historical analyses of reactor construction and operation cost show significant variations amongst different reactors.[23]

Another problem has been a history of time and cost overruns. A recent study of the economic feasibility of nuclear power in Spain came to the conclusion that it "could be competitive only for the nuclear projects with the lowest projected costs… and most importantly, if there are no cost overruns".[24]

The last decade has seen a further decline in the relative economics of nuclear power.[25] Whereas the MIT study assumed a capital cost of $2000/kW as their base case, in the case of the Vogtle project, where two AP1000 reactors are being constructed in the U.S. state of Georgia, a senior official from the utility building the reactor estimated in 2012 that the unit capital costs were $6100/kW, roughly 2.5 times the cost estimate assumed in the MIT study even after allowing for inflation.[26] Since then cost estimates for Vogtle have further increased because of delays and difficulty in meeting quality standards.[27] Likewise, the estimated costs of constructing a European Pressurized Reactor in Western

Europe or North America range from around $5000 to $7300 per kW, or about $6100 on average.[28]

The outlook for the future is not bright either. An expert elicitation study involving 30 U.S. and 30 European nuclear technology experts found that on average, under a business as usual scenario, they expected current (Gen. III/III+) designs to be somewhat more expensive in 2030 than they were in 2010, and they expected the next generation of designs (Gen. IV) to be more expensive still as of 2030.[29]

It is worth emphasizing that KEPCO's deal with the UAE is a special case and one cannot come to any conclusions about the relative economics of nuclear power from this case. The primary aim of the deal with the UAE appears to have been to introduce the South Korean nuclear exports to the region and gain more market confidence. For this reason, South Korea seems to have subsidized the project substantially; some have estimated the deal with the UAE at being about 20 per cent beneath the industry average.[30] Not surprisingly, the deal was criticized within South Korea as commercially weak and that future customers will demand similar terms.[31] While there is a long history of systematic under-bidding in nuclear projects, especially in the case of countries with ambitious nuclear programs, this sort of subsidization can be done only for the first one or two projects, and cannot be the basis of a large-scale expansion of nuclear power in a region.

Cost Analysis

The conventional way to compare the cost of electricity generated by different sources is to calculate the levelized cost of electricity (LCOE).[32] This follows from the standard discounted cash flow methodology, which accounts for the time-value of money.[33] We use this methodology to calculate the life cycle cost of producing electricity. The levelized cost is the ratio of the total cost to the benefits (in this case the electricity produced) with all figures being discounted to the same baseline year.

The calculated costs of electricity are all busbar costs delivered to the grid; i.e., they take into account auxiliary or in-plant consumption of electricity but do not include transmission and distribution costs. Any large-scale expansion of nuclear power or utility-scale solar energy would require an expansion of transmission infrastructure. We do not include those costs here, even though it could be significant.

The cost of electricity generated as well as water produced by desalination from any technology depends on a number of parameters. An important factor is the discount rate. For our base case, we assume a real discount rate (i.e., inflation is implicitly taken into account) of 5 percent. This choice may seem somewhat low, but many studies do indeed adopt discount rates of around 5 to 5.5 percent in their evaluations of electricity economics in the GCC.[34]

The cost of electricity depends not just on the number of years but also the precise distribution of annual capital expenditures. When listed in terms of cumulative expenditure, this distribution is termed an S curve because of its shape. Because of the lack of experience with nuclear construction in the GCC, the simplest distribution—a nearly even distribution of costs, with equal annual expenditures during most of the years except for the first and last year—has been assumed. For natural gas and solar CSP, we assume that the expenditure is equally divided into two years, whereas for solar photovoltaic projects, the expenditure is assumed to be incurred in one year.[35]

A third factor that affects the relative economics of different energy technologies is the capacity factor, the ratio of the energy produced in a given year to the amount that could have been produced if the power plant were to operate at full power all the time. Based on experiences or estimates in the GCC and elsewhere, we assume average capacity factors of 25 percent for solar PV,[36] 35 percent for solar CSP,[37]and 90 percent for nuclear and gas based power.

For capital costs we use figures from the U.S. Department of Energy's Energy Information Administration.[38] The EIA assumes a capital cost estimate for nuclear reactors of $5,530/kW, much lower than the average cost of nuclear reactors under construction.[39] In contrast, the EIA's solar cost estimates are likely too high.[40] Further, the cost of solar power has been dramatically declining. This is particularly relevant because of the long construction period of nuclear projects, a decade at the very least.[41] Even assuming that a nuclear power project is given the go-ahead in 2014, it will likely be 2024 by the time it starts generating electricity.

The estimates of how fast solar costs are expected to decline are striking. In 2012, based on solar industry experts, other research groups, the EIA, and various company web sites, analysts from the McKinsey Consulting Company projected that "the cost of a commercial-scale [solar] rooftop system could be reduced by 40 percent by 2015… and by approximately another 30 percent by 2020—to nearly $1.20 per Wp".[42] In all, therefore, the reduction is about 70 percent relative to 2010. Another study projected roughly 40 percent declines in capital costs for both PV and solar thermal.[43] We use the latter figure, i.e., 40 percent, in capital costs for solar PV and CSP but start with the EIA's current estimate for both costs.[44]

This is a cautious assumption and there is good reason to expect even greater declines in prices. The U.S. Department of Energy (DOE), for example, has initiated the SunShot Initiative, which seeks to make solar energy cost-competitive with other forms of electricity by the end of the decade. It expects to do this by reducing the total installed cost of solar energy systems to $.06 per kWh, or $60 per MWh, by 2020 [45]. By February 2014, the costs had declined sufficiently for DOE to declare that 60 percent of the target had been reached.[46] However, our approach in this paper is to consider assumptions that are relatively favorable to nuclear power. Therefore, for our base case, we only assume a 40 percent decline in solar prices. All of the assumptions used in this analysis are listed in Table 3 below.

The comparison of the levelized cost of electricity between solar CSP, solar PV and nuclear based on 2020 projections is shown in Figure 3. Although nuclear generated electricity is marginally cheaper than that produced by solar, solar technologies have the potential to experience further and rapid decline in costs. If this trend were to continue, then it is very likely that solar electricity would be cheaper than nuclear power well before any nuclear reactor in the GCC, perhaps other than the subsidized Barakah project in the UAE, starts generating electricity.

	Nuclear	Gas	Solar (CSP)	Solar (PV)
Unit Capital Cost ($/kW)	$5,530	$1,023	$5,067	$3,873
O&M cost ratio			1.33%	0.65%
Assumed reduction in capital costs by 2020			40%	40%
2020 Capital Cost ($/kW)			$3,040	$2,324
Fixed O&M ($/kW-y)	$93.28	$15.37	$40.43	$15.10
Variable O&M ($/MWh)	$2.14	$3.27	0	0
Heat rate (BTU/kWh)		6430		
Gas used (mmBTU/MWh)		6.43		
Gas cost ($/mmBTU)		Variable		
LEU Fuel Cost ($/kg)	$3,211		0	0
Fuel consumption (kg/kWh)	3.15E-06			
Fueling costs ($/MWh)	$10.11	Variable	0	0
Economic life	60	40	25	25
Capacity Factor	90%	90%	35%	25%
Auxiliary Consumption	8%	8%	0.25%	0.25%
Discount Rate	5%	5%	5%	5%

Table 3 **Cost Assumptions for Various Technologies**

On the other hand, there is little reason to expect the costs of nuclear power to decline substantially. Indeed, historically, costs of nuclear power have only increased.[47] Therefore, the only variation considered here (Fig. 4) is one where the nuclear construction cost was higher by 30 percent. As for solar technologies, the range of variation in capital cost considered here is between -20 and -40 percent of today's cost of solar technologies. Estimates of solar electricity generation cost are roughly comparable to other values in the literature for utility-scale projects.[48] Further, although we do not consider the possibility of storage, it has been suggested there are latent storage thermal systems that can be developed over the next decade that might be able to produce electricity at below $60/MWh.[49]

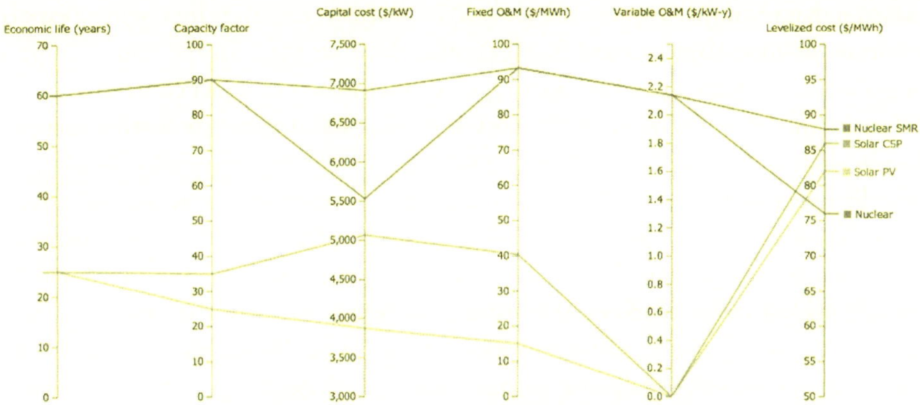

***Figure 3* Levelized Cost of Nuclear and Solar Energy Technologies and Values of Various Input Parameters**

A special case to be considered here is nuclear power from SMRs. As discussed earlier, SMRs are expected to be higher in capital cost per kW by somewhere between 12 and 121 percent.[50] In our analysis, we will assume that the capital cost is about 25 percent higher than for current nuclear reactors, i.e., approximately $6,900/kW. They would also have higher fueling cost because of higher uranium requirements.[51] It is also not clear what sets of conditions they would be licensed under,[52] and depending on the safety and security requirements imposed by regulatory authorities, SMRs could have higher fixed and variable operations & maintenance costs. In contrast to these higher costs, the construction time for SMRs is expected to be shorter.[53] For simplicity, it is assumed that all of the cost variables for current (large) LWRs are the same as SMRs with the exception of the capital cost. Then, with these assumptions, the levelized cost of nuclear power from reactors rises from about $76/MWh if large reactors are constructed to over $88/MWh if SMRs are.

Nuclear versus Natural Gas

Unlike solar power that has zero fueling costs, the cost of natural gas is an important, perhaps the dominant, component of the cost of generating electricity in a natural gas plant. Therefore, the cost comparison between nuclear and natural gas could vary from country to country. In particular, the natural gas prices in the case of countries that use domestic reserves would be significantly lower than countries that import natural gas at international prices, and this would affect the relative economics of power from nuclear reactors and natural gas plants.

Within the GCC, we have three categories: two countries that are natural gas exporters (Qatar and Oman; note that Oman imports natural gas but is a net exporter!);

two countries that are importers (Kuwait and UAE); and two countries that neither export nor import natural gas (Bahrain and Saudi Arabia; there are plans in Bahrain to import natural gas, either through a pipeline from Qatar or via imports of liquefied natural gas[54]). All of these countries would compare the cost of nuclear and natural gas based electricity in different ways.

The difference in the cost of electricity generation between nuclear reactors and natural gas plants decreases as the prices of natural gas increase. The cross-over value between nuclear and natural gas generated electricity is at a natural gas price of $9.5/mmBTU. In other words, nuclear is less economical if natural gas prices are lower than the cross-over value. In the case of the countries that are more suitable for SMRs because of their smaller grid size, the cross-over value of natural gas price below which nuclear power does not make economic sense is $11.4/mmBTU.

The question then is at what prices the GCC countries expect to purchase natural gas over the next few decades. We have examined the case of Saudi Arabia in a previous study and showed that the economic prospects for nuclear power in the kingdom are not favorable in comparison with natural gas, even if the currently low domestic natural gas prices in Saudi Arabia were to rise substantially.[55]

What if Saudi Arabia were to increase its domestic output and start exporting gas instead of using it in natural gas plants? This is related to what is often termed the opportunity cost. Because such exports will most likely be in a liquefied (LNG) form, the expenditures associated with building infrastructure, liquefaction and shipping should be taken into account while evaluating the opportunity cost. A study on the future of natural gas conducted by the Massachusetts Institute of Technology estimates the cost of liquefaction at $2.15/mmBTU, shipping of LNG at $1.25/mmBTU, and regasification at $0.7/mmBTU.[56] In all, these amount to $4.1/mmBTU. Because of this additional expenditure, it would make economical sense for Saudi Arabia to build a nuclear reactor in comparison to a natural gas plant only if the price that could be obtained on the international market exceeds $13.6/mmBTU over the period of the lifetime of the reactor.

For countries that are already exporting natural gas (Qatar and Oman), the shipping and regasification remain the same but the liquefaction costs are expected to be lower given the required infrastructure has already been built. Thus, for them, it would make economic sense to replace natural gas with nuclear power when the price available for natural gas is in the vicinity of $13.4/mmBTU.

While we have listed rather precise figures for natural gas prices at which it makes economical sense to switch to nuclear power, we realize that there are many uncertainties in the many parameters we have used to come up with these figures. Thus, these are two be treated as indicative rather than as precise numbers.

The above costs estimates and comparisons can justifiably be accused of being overly partial towards nuclear power since they omit a number of costs specific to that technology.[57] For example, spent-fuel management and decommissioning costs have been completely excluded from these cost-calculations. We have also not included any provisions for costs associated with dealing with nuclear accidents. Even discounting the impacts on human health, Fukushima has shown how large the financial implications of a catastrophic accident are. Despite advances in nuclear technology, it is still not possible to categorically rule out the possibility of such accidents. It is not clear what kinds of subsidies the GCC governments will extend to nuclear power, and thus we cannot quantify these.

Carbon Costs

Even though GCC states do not have any pricing for carbon emissions so far, it is possible that some form of carbon emission pricing could be introduced over the coming decades, either because of a purely domestic decision or because of a global agreement on carbon pricing. This would impact only natural gas; nuclear power and solar energy are conventionally treated as zero carbon sources.

Advanced and future gas-fired power plants are estimated to emit about 0.4 ton CO_2 eq/MWh.[58] The carbon price, if and when introduced, will likely vary regionally but increase with time. The International Energy Agency assumes between $10 and about $90 per ton of CO_2 in its various scenarios in different regions.[59] We therefore assume $50/t$CO_2$, a figure in the middle of this range.[60] For this carbon price, the crossover point between natural gas and nuclear power is reduced to $8.4/mmBTU instead of the $11.4/mmBTU estimated above without including carbon taxing.

Nuclear Desalination

Besides electricity generation, nuclear power offers the GCC countries the possibility of water desalination to meet growing water demand. Nuclear reactors provide heat with a wide range of temperatures and pressures,[61] and consequently, offer high flexibility and adaptability to different desalination technologies. Moreover, the heat rejected to the environment during normal reactor operation is suitable for some desalination technologies such as the Multiple Effect Distillation (MED) and therefore can be utilized anyway. Nuclear desalination has been proven successful with over 150 reactor-years of experience, mainly in Kazakhstan, India and Japan.[62] Likewise the potential for renewable energies, especially solar energy, has been widely explored.

Desalination can be an important way to deal with varying electricity demands for both nuclear and solar energy. This is because unlike electricity, water can be easily and economically stored. A nuclear power plant that produces desalinated water based on a co-generation model can be adjusted so that it dedicates a large part of its power for water production during off-peak times and switch back to fully generating electricity during peak times. Likewise, solar energy, particularly, through concentrated solar thermal technologies have been used to generate both electricity and desalinated water in Spain.[63]

A main advantage of nuclear desalination is that it has the potential of much larger production capacities than that of gas and solar technologies.[64] Therefore, it benefits from economies of scale and offers the possibility of reduced cost per cubic meter of desalinated water. The comparison between nuclear and natural gas is once again dependent on the prices at which natural gas is traded but also on which desalination technology is chosen. These cross over prices are significantly lower than the cross over in the case of electricity generation. As for the comparison between nuclear and solar CSP and solar PV, nuclear power offers a clear cost advantage particularly when multi-stage flash technology is used.[65]

Conclusion

Nuclear power has been proposed by many as a potential source of electricity generation and desalination for the GCC countries. Of these countries, the UAE has already embarked on constructing four nuclear reactors. Saudi policy makers have advanced ambitious proposals for a rapid buildup of nuclear power capacity over the next two decades. These proposals, however, do not always meet the criterion of economic competitiveness. As we have shown here and in earlier work, nuclear electricity is expected, and well on its way, to become more expensive than that produced by solar technologies in the coming decades. Solar photovoltaic and concentrated solar technologies have both been experiencing dramatic declines in prices whereas nuclear construction costs have remained high. There has also been little or no evidence of decrease in costs associated with learning.

On the other hand, the cost difference between nuclear and natural gas is dependent on the prices at which natural gas is traded in the international market. At low prices for natural gas, nuclear power tends to be more costly when compared to natural gas plants. For natural gas importing countries (the UAE and Kuwait and potentially Bahrain), as long as the purchasing price of natural gas is below $9.5/mmBTU, it would be more cost effective to continue importing natural gas. As for Saudi Arabia, Qatar and Oman, the costs of exporting natural gas shift the cross-over value between nuclear and natural gas to values between $11.4 and $13.6/mmBTU, depending on specific country parameters such as availability of infrastructure.

Countries make energy choices based on multiple factors. Economics plays an important but by no means sole role. Energy security, national prestige, capacity building in high technology areas, and the resultant ability to develop nuclear weapons are all reasons that countries use often to justify acquiring nuclear reactors. Supply side factors also play a role: nuclear reactor vendors have, in the face of limited prospects of sales elsewhere, have tried to make various arguments for why GCC countries should adopt nuclear energy. While we have not explored these factors in this paper, these aspects will likely influence the prospects of nuclear power in the GCC.

Notes

1 Hassan M. Fattah, "Arab Nations Plan to Start Joint Nuclear Energy Program," *The New York Times*, December 11, 2006, http://www.nytimes.com/2006/12/11/world/middleeast/11nuke.html.

2 IAEA, "Power Reactor Information System (PRIS) Database," May 24, 2014, http://www.iaea.org/programmes/a2/.

3 WNA, *Nuclear Power in Saudi Arabia* (London: World Nuclear Association, 2013), http://www.world-nuclear.org/info/Country-Profiles/Countries-O-S/Saudi-Arabia/.

4 George Perkovich, "Nuclear Developments in the GCC: Risks and Trends," in *Gulf Yearbook 2007 - 2008* (Geneva, Switzerland: Gulf Research Center, 2008), 227–38.

5 http://data.worldbank.org/indicator/; http://www.iea.org/

6 There are numerous other designs that are potentially available for sale but we do not list any designs that have so far not been constructed or under construction.

7 Mohamed S. El-Genk, "On the Introduction of Nuclear Power in Middle East Countries: Promise, Strategies, Vision and Challenges," *Energy Conversion and Management* 49 (2008): 2618–28.

8 Charles Ebinger et al., *Models for Aspirant Civil Nuclear Energy Nations in the Middle East* (The Brookings Institution, September 2011).

9 WNN, "Teaming up for Saudi Bids," *World Nuclear News*, September 9, 2013, http://www.world-nuclear-news.org/NN-Teaming_up_for_Saudi_bids-0909137.html.

10 Carina Kamel, "French Aim to Subcontract Saudi Firms in Hope of Nuclear Win," *Al Arabiya*, April 17, 2013, http://ara.tv/vz2qj.

11 A "medium" reactor is one having a power output between 300 and 700MWe.

12 Hard does not mean impossible; various techniques to extract uranium from TRISO fuel have been developed. For example, see Liyang Zhu et al.,"Uranium Extraction from TRISO-Coated Fuel Particles Using Supercritical CO2 Containing Tri-N-Butyl Phosphate," *Journal of Hazardous Materials* 241–42 (November 30, 2012): 456–62, doi:10.1016/j.jhazmat.2012.09.072. Further the United States proposes to reprocess a similar fuel at the Savannah River reprocessing plant.

13 M V Ramana and Zia Mian, "One Size Doesn't Fit All: Design Priorities and Conflicts for Small Modular Reactors," *Energy Research & Social Science*, 2 (2014): 115-124, doi:10.1016/j.erss.2014.04.015.

14 IAEA, *Options to Enhance Proliferation Resistance of Innovative Small and Medium Sized Reactors* (Vienna: International Atomic Energy Agency, 2014), 1.
Note that the IAEA includes reactors of up to 700 MW and uses the term Small and Medium Reactors (again with the acronym SMR) rather than Small Modular Reactors in its deliberations. There is, however, a large overlap in the two categories.

15 National Research Council, *Nuclear Wastes: Technologies for Separations and Transmutation* (Washington, D.C.: National Academy Press, 1996), 421.

16 Mario Carelli et al., "Economic Features of Integral, Modular, Small-to-Medium Size Reactors," *Progress in Nuclear Energy* 52, no. 4 (2010): 403–14; Mario Carelli et al., "Economic Comparison of Different Size Nuclear Reactors," in *2007 International Joint Meeting of the Latin American Section of the American Nuclear Society* (Cancun, 2007).

17 Ahmed Abdulla, Inês Lima Azevedo, and M. Granger Morgan, "Expert Assessments of the Cost of Light Water Small Modular Reactors," *Proceedings of the National Academy of Sciences* 110, no. 24 (2013): 9686–91.

18 The 225 MWe reactor corresponds to the Westinghouse SMR design; since then, Westinghouse has decided to pull back from SMRs. Anya Litvak, "Westinghouse Backs off Small Nuclear Plants," *Pittsburgh Post-Gazette*, February 1, 2014, http://www.post-gazette.com/business/2014/02/02/Westinghouse-backs-off-small-nuclear-plants/stories/201402020074.

19 M V Ramana and Zia Mian, "One Size Doesn't Fit All."

20 Jessica Jewell, "Ready for Nuclear Energy?: An Assessment of Capacities and Motivations for Launching New National Nuclear Power Programs," *Energy Policy* 39, no. 3 (2011): 1041–55.

21 WNN, "Reactors Ready for Floating Plant," *World Nuclear News*, August 7, 2009, http://www.world-nuclear-news.org/NN-Reactors_ready_for_first_floating_plant-0708094.html; Alexandr Nikitin and

Leonid Andreyev, *Floating Nuclear Power Plants* (Oslo: Bellona Foundation, 2011); M. Hadid Subki, "Global Trends, Prospects and Challenges for Innovative SMRs Deployment" (presented at the Long-term Prospects for Nuclear Energy in the Post-Fukushima Era, Seoul (Korea), August 29, 2012), http://www.iaea.org/INPRO/5th_Dialogue_Forum/Wednesday,_29.08.2012/Session_IV_Nuclear_Safety_and_Innovation/2._Hadid_Subki_IAEA_0829_no_distribute.pdf; Yuliang Sun, "HTR Development Status in China" (presented at the 23rd Meeting of the IAEA Technical Working Group on Gas Cooled Reactors (TWG-GCR), Vienna, Austria, March 5, 2013), http://www.iaea.org/NuclearPower/Downloadable/Meetings/2013/2013-03-05-03-07-TWG-NPTD/Day_1/3.Sun.pdf.

22 John Deutch et al., *The Future of Nuclear Power: An Interdisciplinary MIT Study* (Massachusetts Institute of Technology, 2003), 3.

23 Nathan E Hultman and Jonathan G Koomey, "The Risk of Surprise in Energy Technology Costs," *Environmental Research Letters* 2 (2007): 1–6; Jonathan G Koomey and Nathan E Hultman, "A Reactor-Level Analysis of Busbar Costs for US Nuclear Plants, 1970–2005," *Energy Policy* 35 (2007): 5630–42.

24 Pedro Linares and Adela Conchado, "The Economics of New Nuclear Power Plants in Liberalized Electricity Markets," *Energy Economics* 40, Supplement 1 (2013): S123, doi:10.1016/j.eneco.2013.09.007.

25 In the United States, for example, several reactors have been prematurely shut down because they cannot compete with the low natural gas prices Matthew L. Wald, "Nuclear Plants, Old and Uncompetitive, Are Closing Earlier Than Expected," *The New York Times*, June 14, 2013.. A former Nuclear Regulatory Commission has argued that nuclear power has become so uncompetitive that market forces will all but phase out the US nuclear fleet by midcentury Peter A. Bradford, "How to Close the US Nuclear Industry: Do Nothing," *Bulletin of the Atomic Scientists* 69, no. 2 (2013): 12–21, doi:10.1177/0096340213477996.

26 Jeffrey Burleson, "Southern Nuclear's AP1000 Reactors at Plant Vogtle Units 3 & 4" (presented at the Base-load Electricity from Natural Gas and Nuclear Power: The Role of Federal and State Policy, Knoxville, TN, September 20, 2012), bakercenter.utk.edu/wp-content/uploads/2012/10/ppt-6-Burleson-session-I-pdf.pdf.

27 Julie Johnsson, "Scana Reactors Face New Delays and $200 Million Additional Costs," *Bloomberg*, June 5, 2013, http://www.bloomberg.com/news/2013-06-05/scana-reactors-face-new-delays-and-200-million-additional-costs.html. Westinghouse, and the operator, Georgia Power, have sued each other for nearly a billion dollars, with each blaming the other for delays and cost escalations Tom Schoenberg and Julie Johnsson, "Southern Co., Westinghouse Sue Each Other Over Plant Cost," *Bloomberg*, November 5, 2012, http://www.bloomberg.com/news/2012-11-05/southern-co-westinghouse-sue-each-other-over-plant-cost.html.

28 Steve Thomas, *The EPR in Crisis* (London: University of Greenwich, November 2010), http://www.psiru.org/reports/epr-crisis; Phil Chaffee, "Newbuild: Fighting over Risk," *Nuclear Intelligence Weekly*, February 13, 2012; Tim Webb, "Soaring Costs Threaten To Blow Nuclear Plans Apart," *The Times*, May 7, 2012, http://www.thetimes.co.uk/tto/business/industries/utilities/article3406852.ece.

29 Laura D. Anadón et al., "Expert Judgments about RD&D and the Future of Nuclear Energy," *Environmental Science & Technology* 46, no. 21 (2012): 11497–504, doi:10.1021/es300612c.

30 Joshua Chaffin, "Seoul's Nuclear Ambitions Wane," *Financial Times*, April 26, 2011, http://www.ft.com/intl/cms/s/0/0d0122de-7030-11e0-bea7-00144feabdc0.html#axzz2qp4ikcVj.

31 Lee Tae-hoon, "Senior DP Official Says President Lied about UAE Nuclear Deal," *The Korea Times*, February 16, 2011, http://www.koreatimes.co.kr/www/news/nation/2014/05/116_81531.html.

32 We are aware of the pitfalls of using LCOE to compare an intermittent source of power like solar energy with nuclear power. However, as mentioned earlier, this methodology makes solar power look less favorable than if the methodology could incorporate the benefits of the overlap between solar generation and peak demand.

33 Richard A Brealey and Stewart C Myers, *Principles of Corporate Finance* (Boston: Irwin McGraw-Hill, 2000).

34 Yasser Al-Saleh, Paul Upham, and Khaleel Malik, *Renewable Energy Scenarios for the Kingdom of Saudi Arabia*, Working Paper (Norwich, U.K.: Tyndall Centre for Climate Change Research, October 2008);

Center for Engineering Research, *Updated Generation Planning for the Saudi Electricity Sector* (Riyadh, Saudi Arabia: Electricity & Cogeneration Regulatory Authority, March 2006).

35 NEA, *Projected Costs of Generating Electricity* (Paris: Nuclear Energy Agency, OECD, 2010), 40, 58.

36 Shafiqur Rehman, Maher A. Bader, and Said A. Al-Moallem, "Cost of Solar Energy Generated Using PV Panels," *Renewable and Sustainable Energy Reviews* 11, no. 8 (2007): 1843–57, doi:10.1016/j.rser.2006.03.005.

37 Michael Peters et al., "Shedding Light on Solar technologies—A Techno-Economic Assessment and Its Policy Implications," *Energy Policy* 39, no. 10 (2011): 6422–39, doi:10.1016/j.enpol.2011.07.045.

38 EIA, *Updated Capital Cost Estimates for Utility Scale Electricity Generating Plants* (Washington, D. C: Energy Information Administration, Department of Energy, 2013), http://www.eia.gov/forecasts/capitalcost/.

39 The use of data from the EIA implicitly offers a subsidy to nuclear power in the GCC, because the costs associated with nuclear regulation in the United States are spread out over about a hundred nuclear reactors, whereas in the GCC countries they would, at least initially, all spent on a few reactors.

40 Steve Clemmer, "EIA's Analysis of Bingaman Clean Energy Standard Underestimates Role of Renewable Energy," *Union of Concerned Scientists*, May 17, 2012, http://blog.ucsusa.org/eia-analysis-bingaman-clean-energy-standard.

41 According to the IAEA's figures for construction periods, the global weighted average reactor construction time is 96.6 months, or about eight years. See IAEA, *Nuclear Power Reactors in the World: 2013 Edition* (Vienna: International Atomic Energy Agency, 2013). In these figures, the construction period is defined as the time between the "first pouring of concrete to the connection of the unit to the grid". However, some costs that are incurred before the first pour of concrete—for example, money spent on ordering components that take a long time to manufacture. If one were to include those as well, there would be cash flows from a project initiation for about ten years. Note that this could be an under-estimate. In France, the last four reactors commissioned in the late 1990s had an average construction time, between first pour of concrete and grid connection, of 124 months, or more than ten years.

42 Krister Aanesen, Stefania Heck, and Dickon Pinner, "Solar Power: Darkest before Dawn" (McKinsey & Company, 2012), 4, http://www.mckinsey.com/client_service/sustainability/latest_thinking/solar_powers_next_shining.

43 Peters et al., "Shedding Light on Solar technologies—A Techno-Economic Assessment and Its Policy Implications."

44 The cost assumed for PV includes all the other components—for example, blocking diodes, power conditioners and inverters/converters G. K. Singh, "Solar Power Generation by PV (photovoltaic) Technology: A Review," *Energy* 53 (2013): 5, doi:10.1016/j.energy.2013.02.057..

45 Department of Energy, "SunShot Initiative: Mission, Vision, and Goals," *SunShot Initiative*, January 16, 2014, http://www1.eere.energy.gov/solar/sunshot/mission_vision_goals.html.

46 Department of Energy, "U.S. Utility-Scale Solar 60 Percent Towards Cost-Competition Goal," *Energy.gov*, February 12, 2014, http://energy.gov/articles/us-utility-scale-solar-60-percent-towards-cost-competition-goal.

47 Hultman and Koomey, "The Risk of Surprise in Energy Technology Costs"; Arnulf Grubler, "The Costs of the French Nuclear Scale-up: A Case of Negative Learning by Doing," *Energy Policy* 38, no. 9 (2010): 5174–88.

48 Steven Griffiths, "Strategic Considerations for Deployment of Solar Photovoltaics in the Middle East and North Africa," *Energy Strategy Reviews* 2, no. 1 (2013): 125–31, doi:10.1016/j.esr.2012.11.001.

49 K. Nithyanandam and R. Pitchumani, "Cost and Performance Analysis of Concentrating Solar Power Systems with Integrated Latent Thermal Energy Storage," *Energy* 64 (2014): 793–810, doi:10.1016/j.energy.2013.10.095.

50 Ahmed Abdulla, Inês Lima Azevedo, and M. Granger Morgan, "Expert Assessments of the Cost of Light Water Small Modular Reactors," *Proceedings of the National Academy of Sciences* 110, no. 24 (2013): 9686–91.

51 Alexander Glaser, Laura Berzak Hopkins, and M.V. Ramana, "Resource Requirements and Proliferation Risks Associated with Small Modular Reactors," *Nuclear Technology* 184 (2013): 121–29.

52 M.V. Ramana, Laura Berzak Hopkins, and Alexander Glaser, "Licensing Small Modular Reactors," *Energy* 61 (2013): 555–64, doi:10.1016/j.energy.2013.09.010.

53 Abdulla, Azevedo, and Morgan, "Expert Assessments of the Cost of Light Water Small Modular Reactors."

54 EIA, "Bahrain," - *U.S. Energy Information Administration*, February 2013, http://www.eia.gov/countries/country-data.cfm?fips=BA.

55 Ali Ahmad and M. V Ramana, "Too Costly to Matter: Economics of Nuclear Power for Saudi Arabia," *Energy*, 69 (2014): 682–694.

56 Ernest J Moniz, Henry D. Jacoby, and Anthony J. M. Meggs, *The Future of Natural Gas* (Cambridge, MA: Massachusetts Institute of Technology, 2011).

57 For a comprehensive list of subsidies to nuclear power in the United States, see Doug Koplow, *Nuclear Power: Still Not Viable Without Subsidies* (Cambridge, MA: Union of Concerned Scientists, February 2011).

58 Weisser D. A guide to life-cycle greenhouse gas (GHG) emissions from electric supply technologies. Energy [Internet]. 2007 Sep [cited 2014 Mar 8];32(9):1543–59. Available from: http://www.sciencedirect.com/science/article/pii/S036054420700028X

59 IEA. World Energy Outlook 2012. Paris: International Energy Agency; 2012.

60 This carbon price is also in the range of values assumed by large corporations in their internal planning. See Ben Garside, "Big Firms Brace for Global Carbon Price Rollout: Report," *Reuters*, September 15, 2014, http://www.reuters.com/article/2014/09/15/us-carbontrading-global-idUSKBN0HA09W20140915.

61 Note that the heat and pressure of the coolant that come out of the reactor is constant but the heat transferred to secondary systems can be adjusted.

62 WNA, "Nuclear Desalination," *World Nuclear Association*, 2013, http://www.world-nuclear.org/info/Non-Power-Nuclear-Applications/Industry/Nuclear-Desalination/#.UnbKIpTk80M.

63 John Lienhard et al., "Solar Desalination," *Annual Review of Heat Transfer* 15, no. 15 (2012): 277–347, doi:10.1615/AnnualRevHeatTransfer.2012004659.

64 Unsurprisingly, desalination capacities of SMRs would be lower than that of large nuclear reactors due to reduced overall thermal power transfer, with a consequent decline in the financial advantages over natural gas and solar power.

65 MSF works well even with high salinity levels and has a larger production capacity compared to other technologies. For further cost details see Ali Ahmad and M.V. Ramana, "Too Costly to Matter: Economics of Nuclear Power for Saudi Arabia," *Energy*, 69 (2014): 682–694.

5

The Energy Situation in the Middle East: Potential Threats to Oil Export Revenues in Gulf Countries

Yoshikazu Kobayashi

Introduction

In projecting the position and influence of the Gulf countries in the international society, examining the political power of these countries is a necessary step. According to traditional categorization, a country's political power can be broken down into military power, economic power, and power over the opinion.[1] Among the three categories, economic power which is mainly generated from their oil revenue is the very source of Gulf countries' political power. Gulf countries can exercise unique diplomatic influences by such revenues. Gulf countries' sovereign wealth funds, which are mainly funded by oil export revenues, generates additional such influences through investments to various assets and projects abroad. Oil revenues provide essential funds to enable national integration through generous social spending and subsidies. Oil export revenues in Gulf countries itself seem to have peaked in 2012, and uncertainties over the future revenues have been growing (Fig. 1). Examining how these export revenues will flow to the countries and what are the likely threats to the revenues in the future, therefore, will provide a good guide to project the future power of the Gulf counties.

In light of this interest, this paper considers that Gulf countries face three potential threats that may affect the future revenues of oil in Gulf countries, namely, 1) the shale revolution that is underway, centered on North America; 2) the "Arab Spring" that swept through the Middle East region in 2011, and, 3) expanding energy demand and energy shortfalls within the Middle East. The paper then argues that shale revolution will not have a significant impact on the Gulf oil export revenues; but change of crude oil trade pattern and "energy independence" perception in the US may affect geopolitical environment in the Gulf. The "Arab spring" did not cause a large scale supply disruption while raising international crude oil price level; but, situations in Syria and Iraq has a potential to cause a serious disruption. Increased social spending and raised targeted crude oil price caused by rapid demand growth is also a source concern for Gulf countries. Section 1 to 3 discuses these three potential threats, reality and likely impacts to the future revenue amounts, and Section 4 concludes the entire discussions.

The Impact of the Shale Revolution

The factor that first likely to affect oil export revenue of Gulf countries in the future is the shale revolution[2] currently underway chiefly in the US. The US' crude oil production volume has been following a long downward trend since reaching a peak in the early 1970s. However, from around the end of the 2000s, technologies that are applied to shale gas development began being applied to oil development as well. Thanks to these innovative development technologies, particularly a combination of hydro-fracturing and horizontal drilling, oil production from the shale rock increased rapidly, especially in Bakken of North Dakota and the Eagle Ford of Texas. As a result of this progress in developing the shale oil resources, the US' crude oil production volume reportedly increased by over 954,000 barrels per day (b/d) in 2013 compared to the previous year, the largest production increase recorded in the history of the country's oil production.[3] At the same time, domestic oil demand began decreasing in the US from the mid-2000 because of persistent high oil price and improved fuel economy. The country's net oil imports have been falling rapidly from a peak set in 2005, and consequently the idea of oil independence in the US, long viewed as a "pipe dream," is beginning to seem like a realistic goal (Fig. 2).

The shale rock where shale oil resources are located is the source rock where oil and natural gas are formed, and consequently in many cases shale rock exist in locations where there are oil and natural gas fields. Shale resources themselves are therefore distributed widely in regions worldwide, not just in the US (Fig. 3). Work on assessing the overall picture of shale resources is only just beginning and as yet there are no highly accurate statistics on proven reserves, or resources that can be economically produced with existing development technologies. If development and production on the same scale as that being witnessed in the US becomes possible where other shale resources exist, it will lead to a substantial change of the geographical balance of oil resources, and thus has the potential to have a truly "revolutionary" impact to the international oil and gas market.

The rapid growth of US domestic oil production has already caused several phenomena that were not considered to happen before. One of the most evident changes in the international oil market under the shale revolution is an increasing price differential among benchmark prices. As shown in Fig. 4, WTI, or the US benchmark price, has traded at a lower level than Brent, or the European benchmark price, since 2011. This is because the US domestic oil market balance has loosened since domestic production has significantly increased while exporting crude oil is banned in the US. WTI had traditionally traded at a premium to Brent under the assumption that the US always imports crude oil from abroad and any international crude oil has to be economically even to the WTI price factoring in transportation cost. This assumption however was undermined by rapid growth of domestic production and the surplus balance in the US. Because the US domestic crude oil has different quality from Middle Eastern crude oil, this price differential has not cause any notable impact on to the pricing of Middle Eastern crude oil. But, as discussed later, this has indirectly affecting oil product exports from Middle East.

Another phenomenon caused by the shale oil production surge is observed in US crude oil imports. Fig. 5 shows US crude oil imports from major oil exporting countries. Imports from African countries such as Nigeria and Angola have significantly decreased in the recent years. This is because the quality of African crude oil is similar to US domestic crude oil and African crude oil is easily replaced with incremental supply from domestic oil fields. Contrary to the widely perceived view, imports from Saudi Arabia have increased since 2012. This is because Saudi Aramco, the country's national oil company, has a refining joint venture in the US and the refinery expanded its refining capacity in May 2012. Another factor limiting impacts to imports from Gulf countries is crude quality. Because most Gulf countries' crude oils such as Arab light or Arab Heavy, are generally heavier and have higher sulfur content than the typical US domestic crude oil such as Bakken and Eagle Ford (Fig 6), it is more difficult to substitute with US domestic crude oil due to constraints at refining facilities.

It is also notable that US oil product exports continue to increase. The export volume as of 2013 was over 3.4 million barrels per day, and, in fact, the US has become the largest oil product exporter in the world. This surge of oil product export was mainly caused by cheaper crude oil in the US. US domestic crude oil has been traded at a much lower level than the other international benchmark crude oil as mentioned above, and US refiners enjoy the benefits of cheaper feedstock in supplying the international oil products market. Cheaper natural gas supply also helps US refineries by reducing its operating expenses. This increase of US oil product exports has already indirectly affected oil product exports from several export refineries in the Gulf. Some of the US product exports are directed to European markets where export refineries in Gulf countries also ship their products (Fig. 7). In Gulf countries, new export refineries are being constructed and these refineries may need to rethink their planned direction of product exports.

In the US petrochemical sector, investment in ethylene cracking capacity has accelerated to take advantage of cheap ethane feedstocks that have emerged due to the shale revolution. More than 10 million tons of new-build projects are either under construction or consideration in the US. In Gulf countries, the petrochemical industry has been considered a means of industrial diversification and its production capacities continue to expand. Although Gulf petrochemical producers can still maintain their cost competitiveness against US producers thanks to even cheaper ethane feedstock, additional supply from US petrochemical producers will certainly lead to a weaker market balance of petrochemical products in the future.

While these changes were brought on by the shale revolution, at present, few materials exists that can allow us to say that progress in the shale revolution will cause a truly revolutionary impact to the supremacy of Middle East as the lowest cost oil producer. The latest outlook of increases or decreases in oil supply by International Energy Agency going forward to 2030 predicts that Middle East will remain the largest source of supply growth in the long term. Even if progress is made with developing shale resources, it will not alter the superiority enjoyed by the Gulf countries, which possesses abundant reserves that are developable at low cost.

While the shale revolution will not cause a major shift of supply sources in the future, it may bring about major changes in international trade flows. IEE Japan produced a projection for world crude oil trade flows as of 2030 using a global oil trade flow model based on linear programming. Linear programming is a methodology used in operations research to arrive at an optimized solution given provided preconditions. In this analysis, the model calculates the cost minimum trade pattern under the given assumptions of oil price, transportation cost, oil demand in each region of the world, and refining capacities.

According to the model result, as a result of domestic production growth in the US and continued robustness in imports from Latin America and Canada, it will almost be possible to achieve oil independence within the Americas (Fig. 8-b). As of 2030, furthermore, the US may no longer need Middle Eastern crude oil imports. The quality difference issue mentioned above can be solved technically because the US refiners will modify their refining facilities to accommodate the cheaper domestic crude oil. Where export volume from the Middle East to Europe is concerned also, as of 2030, the volume will have declined to 500,000 barrels per day, pressured by declining demand within Europe and increasing supply in Russia and Central Asia. As if replacing this decline in exports to Europe and the US, exports to Asia will meanwhile grow significantly. In particular, it is projected that as of 2030 the Middle East will be supplying China with 6.5 million barrels or more per day and will become China's biggest supply source. This increase in exports of Middle Eastern oil to Asia however is a trend that is already being witnessed. In that sense, the nature of the impact the shale revolution will have on the world's oil trade flows has more to do with further accelerating this existing direction than it does to do with a "revolution."

It is argued that if it becomes possible for the US to stop importing oil produced in the Middle East altogether as a result of the shale revolution, then the US may well curtail its involvement in the Middle East in the future[4]. For the time being, however, at the very least there is little likelihood that the US will drastically revise its policies toward the Middle East. In a nutshell, this is because ensuring the supply of oil from the Middle East represents only a small portion of the US' interests in the Middle East region. Even if US imports of Middle East crude oil were to dwindle to nothing, therefore it would not change in any way the key areas of interest the US possesses in the Middle East outside of oil supply, such as the Middle East peace process and Iran's nuclear development issue, Israel's security, the War on Terror, ensuring crude oil supply for the stable management of the global economy including defending sea lane of communications, and protecting the oil and gas-related interests of US companies in the Middle East.

The US domestic gasoline price will also be affected by the international oil market regardless of whether or not the US will achieve oil independence. This is because, if the domestic gasoline price is cheaper than international market prices, it will lead to gasoline exports to take advantage of the price differentials, and eventually will equalize the domestic and international prices through arbitrage trades. Given the political importance of the domestic gasoline price, the US will need to care about the international oil market balance and therefore will remain committed to Middle Eastern affairs.

Nevertheless, the fact that the US is getting closer to achieving oil independence may potentially be utilized as the perfect "pretext" when pursuing the following types of specific foreign policies. One is the "pivot to Asia" policy the Obama Administration is seeking to advance. In light of the growing importance of Asia to US economic and security interests, the administration established as policy, the position of Asia as a critical area in addition to Middle East in their security strategy.[5] A reduction in the US' dependence on the supply of crude oil from the Middle East also means the elimination of one interest the US has in that region, and when moving ahead with modifying the weight assigned to the Middle East region in US foreign policy, the progress with the shale revolution that is currently taking place will form one basis for that major push.

The next area where shale revolution might affect the US foreign policy is economic sanctions imposed on an oil producing country such as Iran. Under current circumstances, crude oil exports from the US are basically banned, and the US cannot physically make up any lost supply caused by such a sanction. Yet, looked at from a macro perspective growth in crude oil production within the US will lead to an increase in the world's overall oil supply by reducing the US crude oil imports. Consequently, even if tighter sanctions are imposed on crude oil exports by specific oil-producing countries, it will be regarded as possible to compensate for the oil supply lost as a result, and it is in this context that it could be used as an excuse to maintain or tighten sanctions against these specific oil-producing countries. As of writing of this paper (September 2014), in fact, tighter economic sanctions against Russia over its actions against Crimean territory of Ukraine are being discussed in the US. Production growth of shale oil and gas may be mentioned and utilized to justify such sanctions against Russia.

There is also a possibility the US' potential oil independence will become a factor encouraging a so-called Monroe Doctrine policy, whereby the US seeks to lower its external involvement overall. US foreign policy has historically swayed between isolationism and commitment to the international society. Emergence of ample domestic energy supply may strengthen the influence of isolationist views in the US. The US has already achieved self sufficiency in coal and natural gas, and if it is able to achieve independence in its oil supply also the view that the US no longer needs to consider external relations – at the very least where energy supply is concerned – is likely to gain traction. Looking at the US' recent fiscal circumstances furthermore the possibility that US foreign policy will take on an increasingly isolationist tone cannot be ruled out.

Changes Brought about by the "Arab Spring"

The next factor with the potential to have an impact on approaches in the Middle East in the future is the rise of anti-government movements so-called "Arab Spring." This issue is one that may potentially affect the Gulf countries across a broader spectrum than just energy supply and demand. However, this paper will summarize this issue chiefly in the context of oil and energy supply and demand.

An anti-government movement that surfaced in Tunisia in January 2011 subsequently sparked similar movements in Egypt, Libya, Yemen and elsewhere. This had an impact that quite simply upheaved the political situation in the Arab world, including leading to the collapse of decades-long regimes in several countries such as Tunisia, Libya, and Egypt. In the course of that, the Middle East region's oil and natural gas production was of course affected, but the actual adverse effect on supply was not necessarily significant in terms of scale. During the period that this "Arab Spring" swept through the Middle East region, civil movements of some form or other were witnessed in all countries aside from the UAE and Qatar, yet oil and natural gas production was actually affected in only four countries: Libya (oil and natural gas interrupted); Yemen (oil and natural gas interrupted); Egypt (natural gas interrupted); and Oman (oil temporarily interrupted). Of these events, in terms of the impact on the international market, the scale of the interruptions was larger for oil. But even at their peak they reached 1.7 million barrels per day, and when put alongside previous similar cases of supply interruptions, they were not of a scale likely to throw the international oil market into disarray (Fig. 9). Additionally, in the summer of 2011 when the scale of the interruptions was at its largest, the surplus production capacity of OPEC oil-producing countries was around 4 million barrels per day, so the world's oil supply and demand balance was in a position where it was quite capable of absorbing the interruptions.

International crude oil market, however, responded to these anti-government movements in MENA region. After the recovery from the price collapse in the late 2008, international oil market maintained the level around $70/bbl to $80/bbl until 2010. The market rose to the level of $100/bbl and has maintained the level as of September 2014 (Fig. 10). The rise of oil market in 2011 was caused by many factors such as continuous growth of emerging countries' oil demand, quantitative monetary policy by the US, and "Arab Spring" was not the sole reason of price rise. Market concerns over the future political stability of MENA region was undoubtedly one of such causes.

Of the four countries that experienced these oil and natural gas supply interruptions, the largest impacts was observed in Libya. After the Gaddafi Government was toppled by the anti-government movement, Libya recovered to more or less pre-civil war levels in 2012. In the summer of 2013, however, production volume once again fell dramatically due to the impact of political turmoil in the country's eastern region, and its production capacity was also damaged by almost 500,000 b/d of August 2014.[6] In Yemen also, as of August 2014 output remained down by 100,000 barrels per day. Both countries are highly dependent on revenues generated from exporting oil to operate their national economies, and so there are fears of a negative cycle developing in which this unstable supply situation begins affecting their national finances, further destabilizing their domestic political situations.

Despite the fact that Egypt's interruptions did not have a substantial impact on the actual oil and natural gas supply front, the implications that events in Egypt can have on the international oil and natural gas market are considerable because of its strategic position in terms of international trade. As of 2012 crude oil and oil product tankers carrying around 3 million barrels per day were passing through the Suez Canal, and 1.5 billion barrels of oil per day were being transported via the SUMED Pipeline, and Egypt is home to both. When the political situation in Egypt becomes unstable, almost without exception the price of crude oil rises due to fears about flows through the Suez Canal and the pipeline being factored in (even if no actual impact is experienced). In fact, the passage fees that Egypt receives from these trade routes are important sources of foreign currency revenue for the country, and so it has a well-prepared security structure in place for them. The possibility of the operations of this key channel and key piece of infrastructure being impaired as a result of a domestic political conflict is low. However, if for some reason this shipping route and infrastructure were unable to be used, the only option would be the shipping route around the African continent's Cape of Good Hope, and since oil trade would be delayed by the number of additional days required for passage, this would be certain to have a substantial impact on the global oil market.

In addition to the relatively direct and short-term effects of this "Arab Spring," it is conceivable that impacts will be felt in the longer terms in the form of an expansion in these countries' social spending. Following the rise of the string of anti-government movements, all the countries have been attempting to quieten or prevent those movements by boosting domestic social expenditure. Particularly in oil-producing countries in the Middle East, revenue from exporting oil is the main source of finances, and as will be discussed previously, accompanying the countries' growing domestic energy demands the only way of supporting increased social spending such as this is with higher oil prices. Consequently, in oil-producing countries a trend can be detected in which budget breakeven oil prices rise in order to balance fiscal revenue and expenditure across the board. Despite the fact that the Middle East possesses the world's lowest cost oil fields, higher oil prices will become necessary in the future in order to balance the countries' finances, (Fig. 10).

These higher oil prices may be advantageous for Gulf oil-producing countries in terms of securing immediate oil revenue, but in the longer term they will produce side effects. Namely, they will encourage the development of other, relatively high-cost oil fields. The development of shale oil currently underway in the US could be called a perfect example of this. Additionally, higher oil prices will accelerate progress in energy conservation, beginning with the popularization of fuel efficient automobiles and the introduction of alternative energies such as natural gas. This too will have the effect of curbing demand for oil in the long term. The more dependent a country is on high oil prices, the more fragile that country's subsistence base will gradually become, and so the increased social spending witnessed since the rise of the grassroots movements and the growing inclination toward higher oil prices among oil-producing countries that the movements have brought about could invite more serious consequences for Gulf oil-producing countries as well, in the long term.

Changes Brought about by Expanding Energy Demand within the Region

The third factor that can affect oil export revenues by Gulf countries in the future is a growth in domestic energy demand. In recent years energy demand in the Middle East has been growing sharply as a result of population growth and brisk economic activity accompanying sharp rises in crude oil prices. Over the past 10 years energy demand within the Middle East region has grown at an average annual rate of 5.1%, and the quantity of demand has risen by 1.6 times. Even among regions worldwide, this rate of increase is second only to that recorded in the Asia-Pacific region (Fig. 11).

One factor driving this growth in energy demand is that energy prices within the Middle East remain low compared to the international level thanks to government subsidies. As long as those price levels are kept at a constant level regardless of changes in international market conditions, few incentive to conserve energy surfaces among the citizens using it. Fig. 10 shows gasoline prices in the world's key countries, and gasoline prices in countries in the Middle East are clearly low compared to international levels. Fig. 12 presents a breakdown of ratio of subsidies as a component of energy prices in different regions worldwide, and it is clear that even among regions worldwide the Middle East is the region with the highest subsidy rates.

For oil exporting countries in the Gulf, rising domestic demand for energy means a reduction in export volume. Going forward this problem of expanding domestic demand and shrinking export volume will become particularly pronounced in Saudi Arabia, a country with a large population where domestic oil demand is rising sharply.

Oil is usually used as energy for transportation, such as gasoline and diesel for automobiles, jet fuel for airplanes, and bunker fuel for ships. In Saudi Arabia, however, in addition to this transportation-related demand, large quantities of oil are also being used for power generation (Fig. 13). Previously in Saudi Arabia natural gas was used as the main fuel for power generation, but in recent years domestic demand for electricity has grown at a faster pace than growth in domestic natural gas production volume. As a result, large quantities of oil have come to be used as a back-up power-generation fuel.

It appears certain that power generation-related oil demand in Saudi Arabia will grow alongside its transportation-related oil demand. There are even forecasts surfacing that if Saudi Arabia's crude oil production capacity remains constant at current levels and domestic demand alone continues to grow, then by 2040, the country will end up becoming a net importer of oil.[7] In order to avoid such a scenario, investment in exploration toward boosting future natural gas production to meet the domestic energy demand growth is taking place in Saudi Arabia. In its Eastern Province, Saudi Aramco invited European, Russian, and Chinese oil companies to explore natural gas fields[8]. The company is also engaging in natural gas development in Eastern Province and Red Sea by itself, and Karan natural gas field in Eastern Province in fact commenced commercial production in 2011.

To a greater or lesser extent, countries in the Middle East will all face the problem of growing domestic energy demand. As a way of dealing with this problem some countries have begun embarking on promoting energy conservation and the development of alternative energy. For example, in the UAE policies for promoting energy conservation are being prepared that include setting electricity charges according to the amount used and a labeling system for buildings and energy-saving devices. Saudi Arabia also founded the Saudi Arabia Energy Efficiency Center (SEEC) in October 2010 in the King Abudallah City of Science and Technology. The center aims to prepare for comprehensive energy efficiency improvement policies from a labeling system for energy efficiency appliances, and education for efficient use of energy, to development of energy consumption statistics to measure the degree of efficiency of energy consumption. In the industrial sector of the country, Saudi Aramco has also seriously engaged in efficiency improvements in its operations. The company's annual report for the year 2013 revealed that it has improved its energy intensity, or the energy consumption to produce a unit of economic value added, by 2% annually since the year 2000. The report also suggests that the amount of energy saved is equal to 150,000 barrels of oil equivalent per day.

However, in countries in the Gulf, actual energy conservation outcomes have so far been limited. Energy conservation and efficiency improvement initiatives mentioned above have appeared in only a handful of examples. In order to make further real progress in energy conservation efforts, "hard measures," or setting up a specialized government organization will not be sufficient. "soft measures" such as provision of a regulatory framework for energy conservation, identifying benchmarks and setting stretching but realistic targets for energy conservation, developing a regular monitoring system, educating specialists that have an extensive knowledge about energy conservation are also required. Above all, creating a public consensus that energy resources are precious and need to be consumed as efficiently as possible is the most important but challenging task. In Gulf countries, as mentioned above, because domestic energy prices are kept low, and also because citizens have grown accustomed to the energy-intensive lifestyles they have enjoyed over a long period of time, promoting such an idea is not easy.

Where alternative energy initiatives are concerned, the introduction of renewable energy is being explored. King Abdullah City for Atomic and Renewable Energy (KACARE), which Saudi Arabia newly established in April 2010, is promoting the introduction of renewable energy domestically. Saudi Arabia has set a goal of lifting its power-generating capacity from 55 gigawatts (GW) as of 2012 to 120 GW by 2020, and it plans to have renewable energy sources cover 10% of that. In the UAE also, in January 2008 the Abu Dhabi Emirate launched the so-called Masdar Plan for introducing renewable energy, and it is pursuing an economic development program for promoting a sustainable society, central to which are advanced energy technologies (sunlight and solar heat, wind power, hydrogen and CCS (the subterranean storage of CO_2)). The emirate has presented a plan to cover 25% of its electricity demand using nuclear power generation and 7% using renewable energy, centering on solar energy, by 2020. Nevertheless, endeavors such as this are also facing problems, including budget shortages and delays in communication between state-run organizations, and at present things are not proceeding as smoothly as initially planned.

Another alternative energy promotion measure is the introduction of nuclear power generation. Nuclear power plants do not use oil or natural gas as fuel and make it possible to secure large quantities of electricity, and based on such advantages they are undergoing consideration by countries in the Middle East. Iran already operates the Bushehr Nuclear Power Plant, and a large number of plans to introduce nuclear power plants are being pursued elsewhere in the Middle East region, including by the UAE, which is in the process of constructing a nuclear power plant with a 1,200-MW capacity that it aims to start operating in 2017. However, the adoption of nuclear power is accompanied by many challenges, including dealing with the nuclear non-proliferation issue and the problem of disposing of spent nuclear fuel, and so introducing nuclear power plants will not on its own be enough to resolve the Gulf's energy shortfalls.

Conclusion

This paper cited three factors as potential threats that may affect the future oil export revenue of Gulf countries. All of the three factors share a common aspect, namely they will have either few or limited impacts in the short-term but will have a more serious and significant impact in the long-term. Among all the factors mentioned, acceleration of domestic energy demand growth will be the factor that is most likely to occur regardless of developments in the political and economic environments of Gulf countries.

What can Gulf countries do to cope with these long-term threat scenarios? The most effective and necessary action will be to accelerate the current efforts for energy conservation. Gulf countries will be able to allocate more oil for export through reduction of domestic oil consumption. Such conservation efforts will contribute to reduce energy subsidies and thus to ease the pressure to fiscal balances.

In this regard, Gulf countries may start with something feasible, by introducing energy efficiency standards in public sector buildings, or energy efficiency standards to the public procurement process. These actions may not have a notable effect in easing demand growth, but would be a good catalyst to promote an idea that energy conservation is a nation-wide challenge to maintain the current economic power in each Gulf country.

One another effective action is benchmark management. Energy intensity, or energy consumption per unit of GDP, will be a meaningful and useful benchmark. It is needless to say that the goal of energy conservation is to reduce the amount of energy consumption. Energy consumption however tends to increase as economic growth and the two factors often conflict. Energy conservation can be broken down into three factors, namely energy intensity, GDP per capita, and population.

$$\text{Energy consumption} = \frac{\text{Energy consumption}}{\text{Economic output (GDP)}} \times \frac{\text{Economic output (GDP)}}{\text{Population}} \times \text{Population}$$

Energy intensity ***People's living standard***

Because controlling the level of living standard and the number of children is practically difficult, focusing on energy intensity through regular monitoring is a realistic and meaningful action.

Energy conservation usually requires unpopular policy for people because it will accompany subsidy reform. It also required a long time span to realize its effect. But, thanks high oil prices since 2011 people's income in Gulf countries are high enough to absorb subsidies cut. The governments also have sufficient fiscal and political capital to pursue such efforts. It will be more painful for people to accept any economic reform when the Gulf countries economies are not good. Demanding policy actions under the economically difficult time may trigger another public discontent against the government and may turn into an uncertain factor for Gulf region. The age of ever increasing oil export revenues for Gulf countries is certainly ending. Serious and immediate actions are required to Gulf countries.

Appendix

Figure 1 Oil Export Revenues in Selected Oil Producing Countries

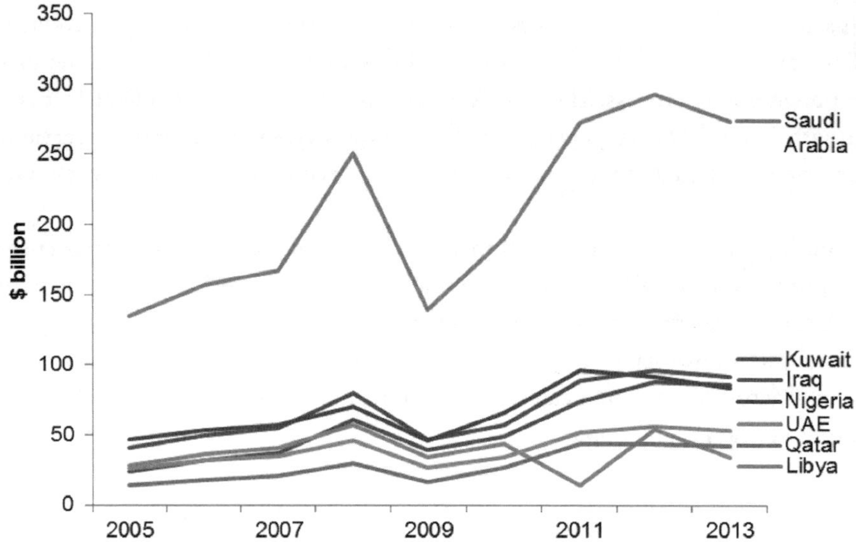

Source: OPEC Annual Statistical Bulletin.

Figure 2 US Historical Net Oil Imports

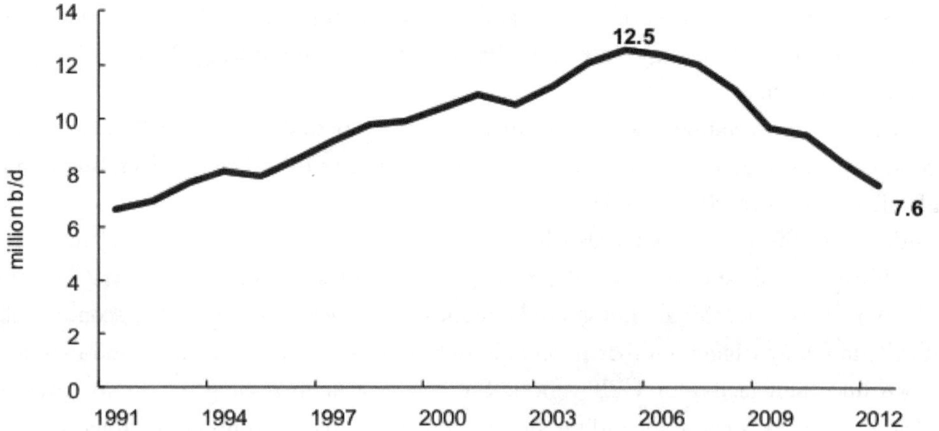

Source: *US Energy Information Administration website.*

Figure 3 The World's Shale Resources

Figure 1. Map of basins with assessed shale oil and shale gas formations, as of May 2013

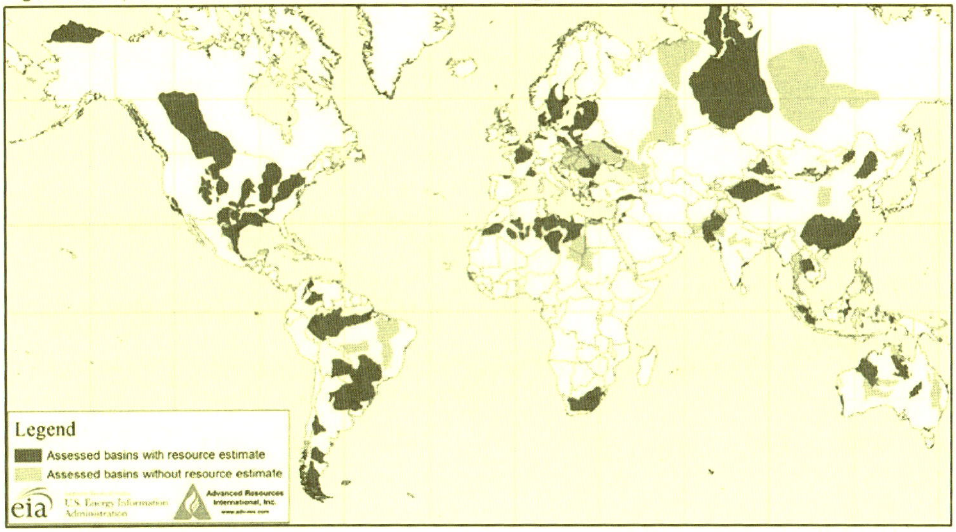

Source: United States basins from U.S. Energy Information Administration and United States Geological Survey; other basins from ARI based on data from various published studies.

Source: US Energy Information Administration website.

Figure 4 International Crude Oil Prices

Source: US Energy Information Administration website.

Figure 5 **Crude Oil Imports from Major Oil Exporting Countries**

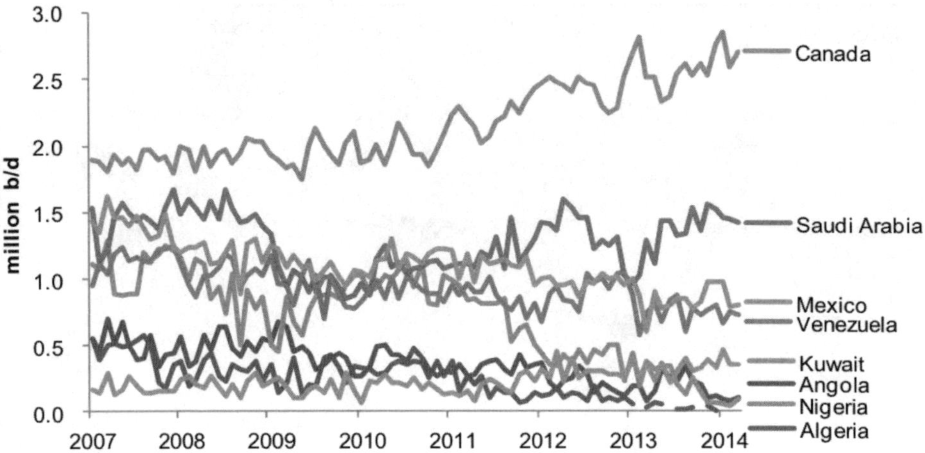

Source: US Energy Information Administration website.

Figure 6 **Quality of Selected Crude Oil Grade**

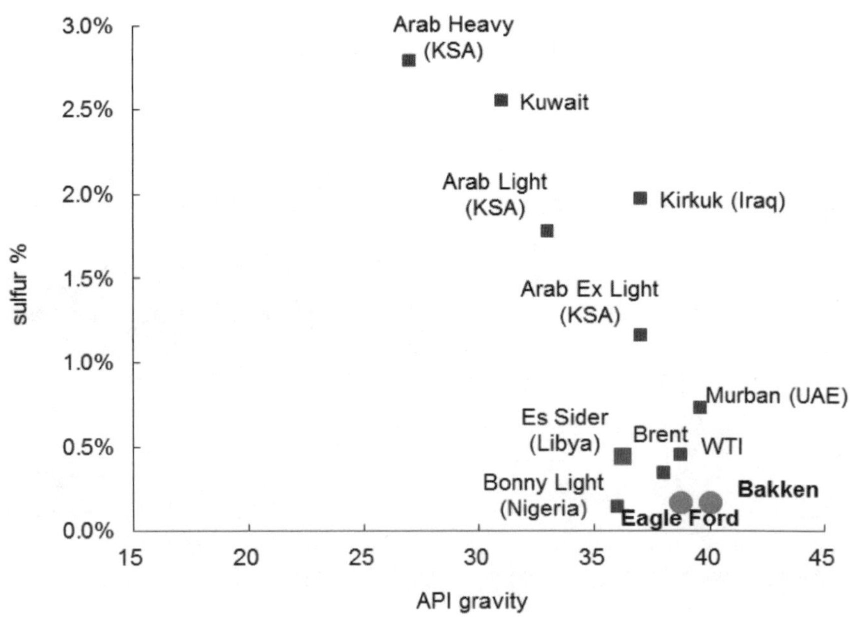

Source: Energy Intelligence Group, International Crude Oil Market Handbook.

Figure 7 Oil Product Exports from the US

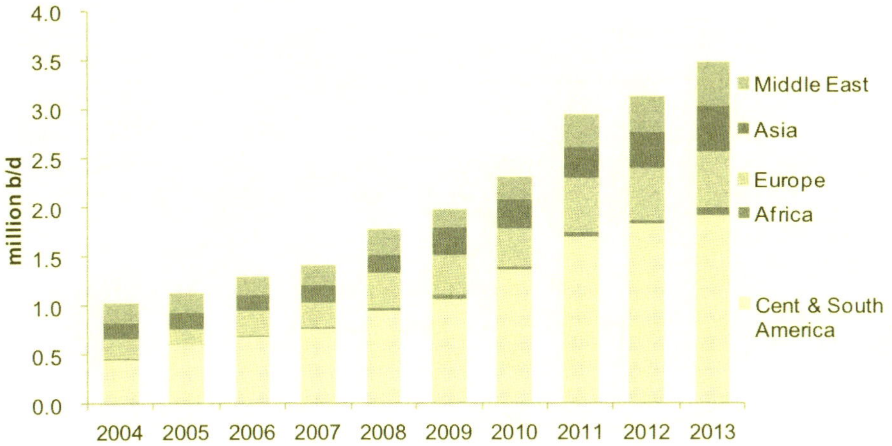

Source: US Energy Information Administration website.

Figure 8-a The World's Crude Oil Trade Flows as of 2012

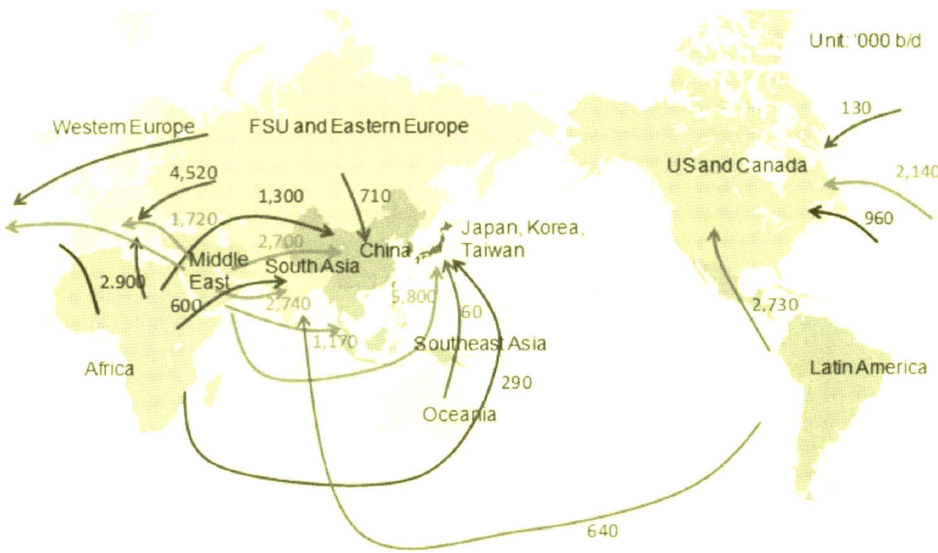

Source: IEE Japan.

Figure 8-b Projections for the World's Crude Oil Trade Flows as of 2030

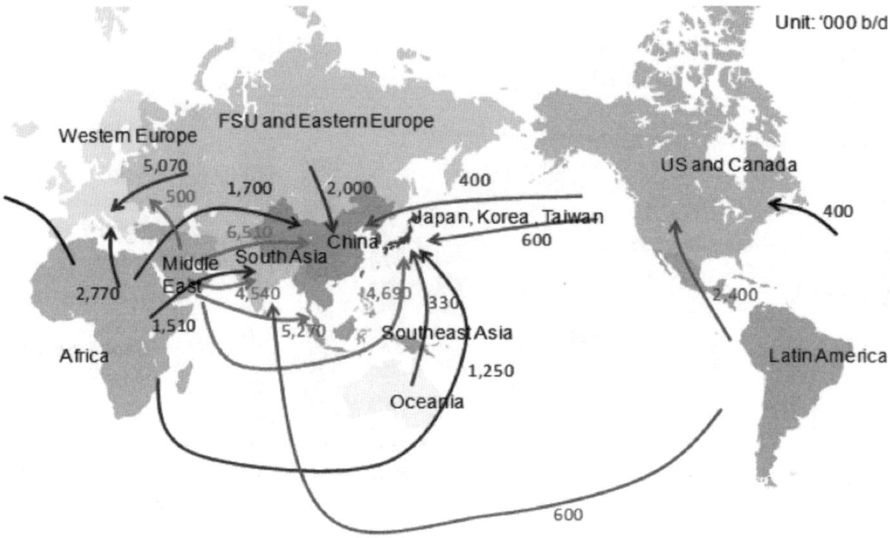

Source: IEE Japan.

Figure 9 Historical Oil Supply Disruptions since the 1970s

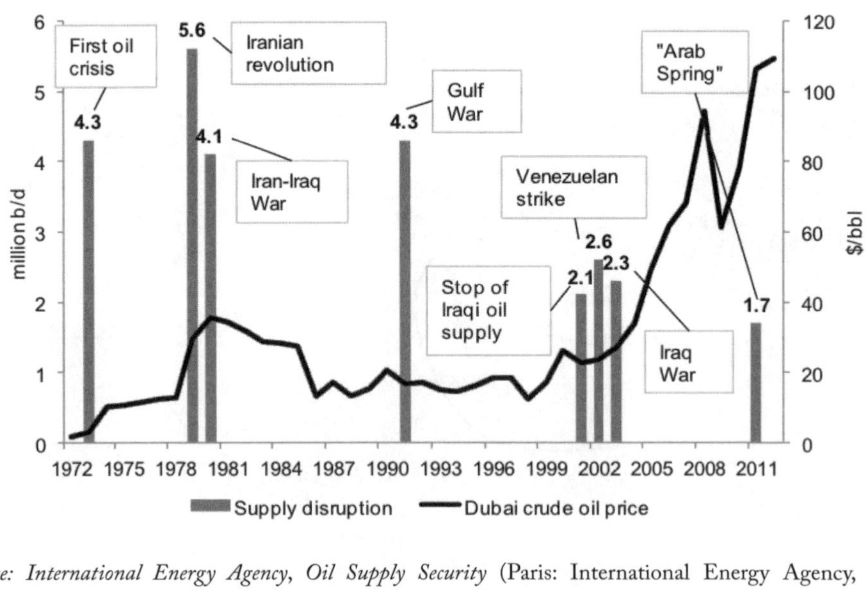

Source: International Energy Agency, Oil Supply Security (Paris: International Energy Agency, 2007). Added by Author.

Figure 10 **Budget Breakeven Oil Prices and Crude Oil Supply Costs**

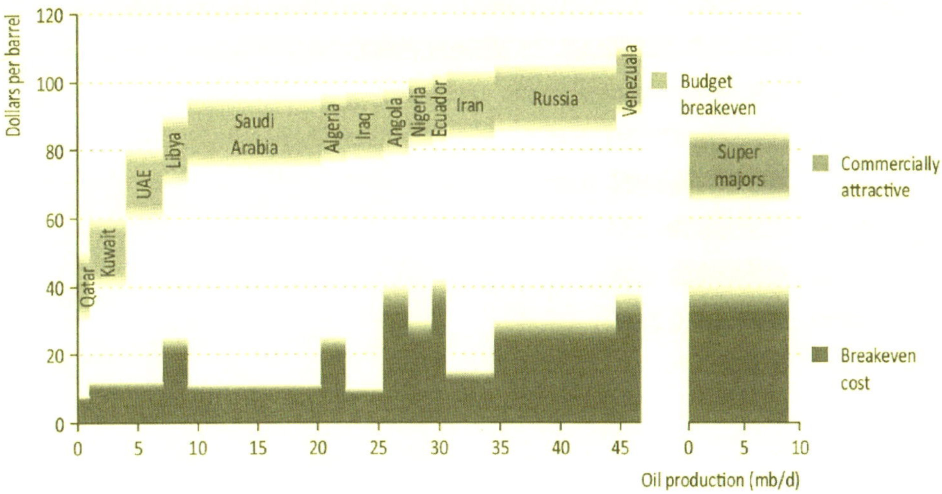

Source: International Energy Agency, World Energy Outlook (Paris: International Energy Agency, 2013).

Figure 11 **Rate of Growth in Energy Demand over the past 10 years**

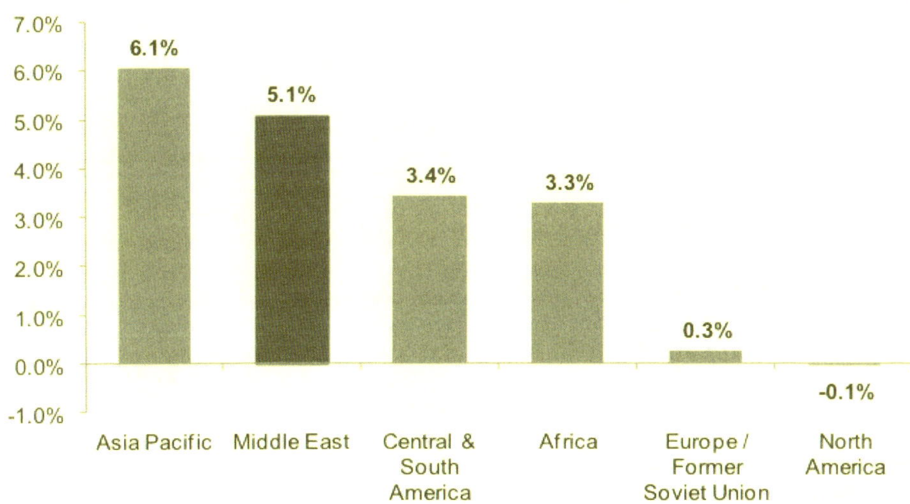

Source: BP Statistical Review of World Energy 2013.

Figure 12 Gasoline Prices by Country (as of 2010)

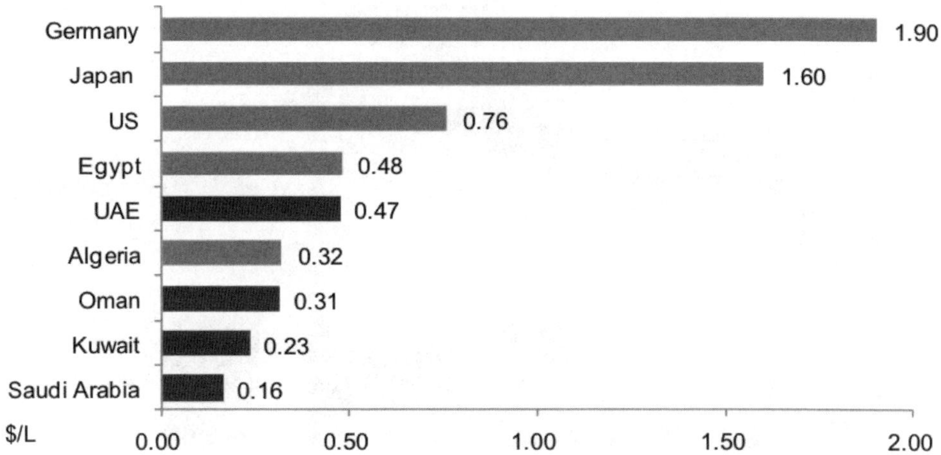

Note: Prices of Germany, Japan, and the US includes taxes.

Source: World Bank, World Development Indicators (Washington D.C.: World Bank, 2012).

Figure 13 Energy Subsidy Provision Rates by Region

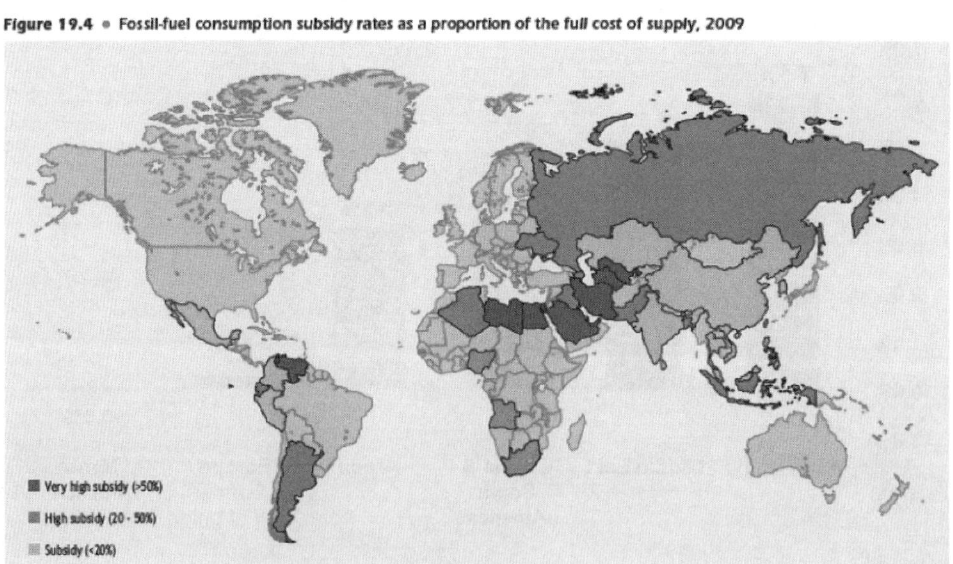

Source: International Energy Agency, World Energy Outlook (Paris: International Energy Agency, 2013).

Figure 14 Saudi Arabia's Oil Demand by Usage

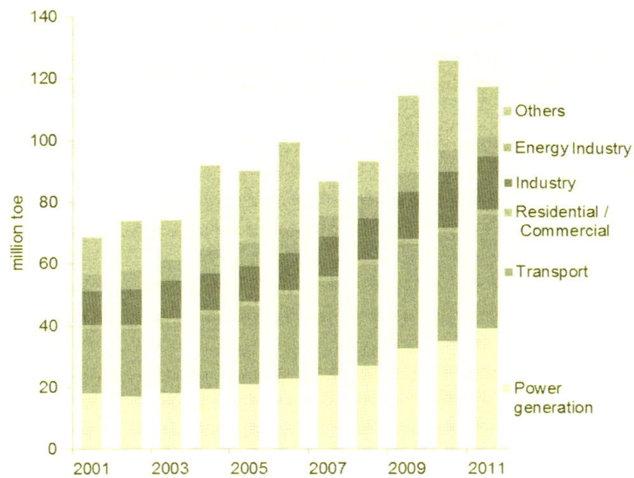

Source: International Energy Agency, Energy Balances of non-OECD Countries (Paris: International Energy Agency, Various editions).

Figure 15 Suez Canal and SUMED Pipeline Flows

Source: US Energy Information Administration website.

Notes

1 Edward H. Carr, *Twenty Years' Crisis, 1919-1939: An Introduction to the Study of International Relation* (New York: Perennial, 2001).

2 Accompanying progress with developmental technologies, in the mid-2000s it became possible to recover oil and natural gas from tight strata, such as shale stratum, that had been prohibitively expensive to develop up to then, and that output is growing rapidly. This phenomenon is referred to as the shale revolution.

3 Energy Information Administration website, available from http://www.eia.gov/dnav/pet/pet_crd_crpdn_adc_mbblpd_a.htm, accessed on 10 September 2014.

4 Roger Howard, "How Shale Energy Reshapes American Security," *National Interest*, May 2013.

5 US Department of Defense, Sustaining U.S. Global Leadership: Priorities for 21st Century Defense, Washington D.C.: U.S. Department of Defense, January 2012.

6 International Energy Agency, *Oil Market Report* (Paris: International Energy Agency, May 2014).

7 Glada Lahn and Paul Stevens, *Burning Oil to Keep Cool: The Hidden Energy Crisis in Saudi Arabia* (London: Chatham House, 2011).

8 Foreign companies that are operating in exploration as of May 2014 are, Sinopec (China), Shell (UK/the Neatherland), LUKOil (Russia), ENI (Italy), and Repsol (Spain).

References

(1) BP, *BP Statistical Review of World Energy* (London: BP, 2013).

(2) Howard, Roger, "How Shale Energy Reshapes American Security," *National Interest*, May 2013.

(3) International Energy Agency, *Energy Balances of non-OECD Countries 2013 edition*, (Paris: International Energy Agency, 2013).

(4) International Energy Agency, *Oil Market Report*, (Paris: International Energy Agency, August 2011, September 2011, October 2011, and May 2014).

(5) International Energy Agency, *Oil Supply Security: Emergency Response of IEA Countries*, (Paris: International Energy Agency, 2007).

(6) International Energy Agency, *World Energy Outlook 2011*, (Paris: International Energy Agency, 2011).

(7) Edward H. Carr, Twenty Years' Crisis, 1919-1939: An Introduction to the Study of International Relation (New York: Perennial, 2001).

(8) Glada Lahn and Paul Stevens, *Burning Oil to Keep Cool: The Hidden Energy Crisis in Saudi Arabia* (London: Chatham House, 2011).

(9) US Energy Information Administration web-site < http://www.eia.gov/oil/>.

(10) World Bank, *World Development Indicators 2012*, (Washington D.C.: World Bank, 2012).

6

Saudi Arabia's Nuclear Energy Plans: Progress, Prospects and Politics

Sujata Ashwarya Cheema

Introduction

Saudi Arabia is seeking to harness nuclear energy through an expansive programme that would meet, albeit partially, the kingdom's soaring energy requirements. Sharply rising demand for power spurred by population growth, a rapidly expanding industrial sector led by petrochemical cities, huge water desalination plants, high demand for air conditioning during the peak summer months, and heavily subsidized electricity disbursal, have put upward pressure on the country's fossil fuel resources. Power generation in the Kingdom of Saudi Arabia (KSA) – as in the rest of the Middle East in general – is primarily thermal, powered by hydrocarbons such as crude oil and fuel oil, enabled by one of the world's largest hydrocarbon reserves and historically high oil prices.

Spiralling domestic demand for electricity prompted a vast power generation expansion programme in 2010, as the KSA plans to double the installed generating capacity by 2032. The kingdom anticipates a rise in domestic oil consumption, currently about 3 million barrels per day (mb/d), to around 8 mb/d, approaching its current output in 2030.[1] Although Saudi Arabia has the world's largest proven oil reserves in the world,[2] capable of supplying crude for the next 80 years at current production level,[3] the anticipated rise in domestic energy demand will start restricting the export capacity of the country. Petroleum export, which accounts for 90 percent of government income,[4] acquires renewed salience in the context of the Arab Spring. The Saudis need to keep up their large social welfare spending for many years to come, so as to prevent the unrest that has engulfed several countries in the region from spilling over into the kingdom.

Thus the world's leading oil exporter has laid the groundwork for ambitious non-conventional energy plans in an effort to curtail the country's dependence on fossil fuels and free up more oil for export. By 2032, Saudi Arabia strives to generate almost 55 gigawatt (GW) of power from non-conventional sources, about 40 percent of the country's power requirements. The kingdom is seeking a diversity of solutions among non-conventional energy or renewable energy – namely, wind, solar, geothermal, waste, and nuclear. In June 2012, the King Abdullah City for Atomic and Renewable Energy (KACARE), the kingdom's energy diversification establishment, announced that Saudi Arabia plans to include nuclear power in its repertoire of non-conventional energy generation.

The KSA has signed nuclear cooperation agreements with France, China, South Korea and Argentina, enabling it to purchase technology and equipment from these countries. Treaties with Russia, Britain, Hungary, Finland, Japan and the United States are under discussion. Saudi Arabia is preparing to call for preliminary bids for its first nuclear reactor, which would inaugurate the region's largest atomic energy programme, a much-needed lifeline – it is believed – for the global nuclear industry fraught by post-Fukushima public cynicism and the global economic meltdown.

Although Saudi Arabia has vocally supported and reiterated the establishment of a nuclear-weapons-free zone (NWFZ) in the Middle East, the kingdom's quest for nuclear energy has not been without its share of controversies. Many fear that once the KSA has a nuclear energy programme, it could develop nuclear weapons either indigenously or with external assistance. A prevalent perception is that it is possible to do so – especially as Iran seeks to master the critical technologies in the nuclear fuel cycle – even as a variety of factors militate against such an effort. Secondly, a periodically surfacing allegation relates to Saudi funding for Pakistan's nuclear weapons programme since 1975. It is rumoured that Riyadh views Pakistan's 'Islamic bomb' as contributing to its own security and that also accounts for the kingdom's support to Islamabad's missile development programme. Some are of the opinion that under a 'security arrangement', Saudi Arabia can summon nuclear weapons from Pakistan for deployment on its own soil in a dire security situation.

While such allegations and fears are largely unsubstantiated by any concrete document in the public realm, the recent surge in interest on the subject of nuclear energy in Saudi Arabia (and indeed other Gulf Cooperation Council, GCC, countries) appear to represent some kind of a reaction to Iran's controversial nuclear programme that many in the West believe, has a potential weapons development facet. While both nuclear energy programme and nuclear weapons production follow similar technical path, any country's decision to pursue the latter is always a conscious political decision. Saudi Arabia, which has signed the Nuclear Non-Proliferation Treaty (NPT) in 1988 and the comprehensive safeguards agreement with the International Atomic Energy Agency (IAEA) in 2005, will certainly weigh its cost in terms of international sanctions and repercussions on the strategic US-Saudi relations, if it were to consider a nuclear weapons programme in the unspecified future.

In the background of the above, this paper argues that:

a) Saudi Arabia soring domestic consumption of oil and gas, maintained by governmental largesse that is difficult to reduce in the context of Arab uprisings, has created an urgent need for supply side energy management, so that the nation can free-up more oil for export. The Saudi oil has two critical value: large production volumes stabilises global supply and price equilibrium; and the revenue earned from oil sale is critical to the kingdom's inflated budgetary spending, crucial to maintaining the political status quo in the context of the regional turmoil. Therefore, diversifying its energy mix in the form of renewable and nuclear energy is a compelling economic and political need for the KSA. Nuclear power is an important segment of the solution envisaged by the kingdom.

b) While any talk of civilian nuclear programme immediately conjures the alarming picture of its transmutation into weapons programme, several critical factors militate against Saudi Arabia pursuing the nuclear weapons path, not least political. Saudi Arabia lacks the natural resources, technological capability, and scientific community necessary to develop a nuclear weapons programme. It has precious little experience in several basic aspects of the nuclear fuel cycle, such as, enriching uranium for reactor fuel, nuclear fuel conversion, nuclear fuel fabrication, and operating nuclear reactors, that must be mastered to develop nuclear weapons.

c) Finally, if Saudi Arabia perceives a fraying of its strong relations with the United States and consequently feels vulnerable about its security, it may consider a nuclear option, either by acquiring it from outside or developing an indigenous programme, the technological limitation for the latter notwithstanding. However, this seems to be a renegade's choice of action, which will expose the kingdom to several vulnerabilities, even as the Saudi state faces long-term insecurities vis-à-vis Iraq, Syria and Iran. The only case in which the Saudis will be compellingly motivated to go nuclear is a scenario in which Iran develops nuclear weapons and this seems unlikely with the prospects of a nuclear deal between Iran and the world powers, in which several aspects of the latter's nuclear programme will be curtailed.

This paper is divided into three major sections. The first deals with the high energy consumption patterns in Saudi Arabia, its potential impact on international oil market and domestic revenue earnings of the kingdom, and the response of the Saudi policy makers to manage internal demand growth of the critical oil resource, once such policy outcome is the development of nuclear energy. Saudi Arabia has taken steps to put institutional mechanisms in place for launching a civilian nuclear energy programme. It has signed agreements with several countries for technical collaboration in the field. These aspects constitute the second section of this paper. The third section looks at the worries and apprehensions caused by Saudi Arabia's nuclear energy programme, especially the concern of its transmutation into a weapons programme, and the conditions under which that could possibly happen.

The Saudi Energy Challenge: Consumption, Consequences and Responses

International observers are keenly watching the skyrocketing domestic energy consumption in Saudi Arabia – where supply relies exclusively on oil and gas – for its critical implications to the global energy scenario. Saudi Arabia's total energy consumption of 6.5 toe (tonnes of oil equivalent)/capita is more than 3 times the world average of 1.9 toe. As the world's 12th largest consumer of primary energy, oil comprised 135 million tonnes of oil equivalent (mtoe) and natural gas 93 mtoe, out of the total consumption of 228 mtoe in 2013. Oil and gas, thus, constituted 59 percent and 41 percent respectively in the kingdom's energy mix.[5] Energy consumption in the country is growing faster than gross domestic product (GDP), leading to a rise in the energy intensity of growth, a general trend observed in most developing countries. Energy-intensive industries as well as in maintaining energy-intensive lifestyles – both encouraged by large fuel subsidies and low electricity prices[6] are consistently driving demand on the upward trail.

In comparison with Qatar and UAE, whose energy supply comes primarily from natural gas, Saudi Arabia's oil consumption appears even starker especially as its proven natural gas reserves stand at 288 trillion cubic feet (tcf), the fifth-largest natural gas reserves in the world, according to the Energy Information Agency (EIA) estimates. A majority of Saudi gas fields is associated with petroleum deposits, or found in the same wells as crude oil. Unlike Qatar, Saudi Arabia does not export natural gas – the entire production of 4 tcf, is used up for domestic consumption.[7] Saudi Aramco, the state's oil firm, forecasts that the natural gas demands in the kingdom is expected to almost double by 2030. All current and future gas supplies (except natural gas liquids) are now earmarked for meeting domestic demand in order to lessen the burdensome reliance on oil.

Consumption Patterns and Demand Drivers

The surge in Saudi energy consumption in the past decade, from 136 mtoe in 2003 to 228 mtoe in 2013[8], amid dominant reliance on petroleum liquids throughout the ten-year period (Table 1) and anticipated further growth in consumption, has raised serious concerns about Saudi Arabia's capacity to export large volumes of oil in the future and remain the traditional bulwark of secure supplies. In 2013, Saudi Arabia consumed 135 million tonnes of oil, a significant 25 percent out of its total production of 542 million tonnes.[9] In fact, Saudi domestic demand for oil shows a steady ascendance since 2008 (Table 2).

Domestic energy consumption is led by summer surges in demand for air conditioning, which is 40 percent higher than in winters. Base and peak load demands also fluctuate widely during day and night requires the total installed capacity to be considerably above base load requirements.[10] The people of Saudi Arabia use on average nine times more electricity than their fellow Arab counterparts in Egypt, Algeria, Sudan and Morocco, according to a report published by the economic reporting unit of *Al-Eqtisadiah* newspaper. It states that an individual in Saudi Arabia consumed on average 8,161 kWh in 2011, compared to 951 kWh by individuals in the four largest Arab countries. The household sector consumes 50 percent of the Kingdom's total electricity production.[11]

Table 1 **Prevailing Reliance on Oil in the Primary Energy Mix (2003-13) in million tonnes of oil equivalent (mtoe)**

Year	Total Primary Energy Consumption (TPEC)	Oil Consumption	Percentage of TPEC	Natural gas Consumption	Percentage of TPEC
2003	136	82	60%	54	40%
2004	147	88	60%	59	40%
2005	152	88	58%	64	42%
2006	158	92	58%	66	42%
2007	165	98	59%	67	41%
2008	179	107	60%	72	40%
2009	186	116	62%	71	38%
2010	203	124	61%	79	39%
2011	208	125	60%	83	40%
2012	220	131	60%	89	40%
2013	227	135	59%	93	41%

Source: BP Statistical Review of World Energy, June 2014.

Table 2 **Saudi Arabia's Year-wise Oil Production, Consumption & Consumption as % of Production, 2003-13 (million tonnes)**

Year	2003	2004	2005	2006	2007	2008	2009	2010	2011	2102	2013
Production	486	500	521	509	489	510	457	474	526	550	542
Consumption	82	88	88	92	98	107	116	124	125	131	135
Consumption as % of production	17%	18%	17%	18%	20%	21%	25%	26%	24%	24%	25%

Source: BP Statistical Review of World Energy, June 2014.

Electricity use is increasing at about 7.5 percent annually, causing the overall energy demand to rise nationwide. Peak demand for electricity is obviously, higher, experiencing an increase between 8-10 percent a year. Fuel oil and diesel provide around 57 percent of Saudi electricity, with natural gas providing the rest. In recent years more and more crude oil has been diverted to the power sector. Since 2011, more than 500,000 barrels or 500 kilo barrels per day (kb/d) are burned for power generation, with peak demand in summer now seen topping 900 kb/d. Demand for electricity is already running up against the kingdom's installed power generation capacity of 58.4 GW as of 2013.[12] The demand growth in electricity has already exceeded the capacity of the gas and fuel oil sectors to supply the power stations, such that the Saudi Aramco imported one of the largest quantities of fuel oil in more than a decade in 2013.[13]

In oil rich economies such as Saudi Arabia, revenue earned from sale of oil not only finances public spending but also plays a determining role in economic diversification.[14] Saudi policy makers' drive to diversify the kingdom's oil-reliant, mono sectoral economy reveals the dilemma of reconciling greater hydrocarbon exports to economic diversification. Development of non-oil sectors, such as, manufacturing and construction as well as industries, such as, aluminium and steel will drive greater demand for energy. While energy from oil (and gas) would be required run a diversified economy, the same energy is also needed for export to earn revenue for governmental efforts at diversification. This predicament plays out no less in the diversification of the oil sector. Saudi Arabia has exploited its comparative advantage of sizeable oil reserves to produce downstream products such as chemicals, fertilizers, and petrochemicals. Gas shortages have encouraged oil substitution in the petrochemical industry putting further pressure on Saudi oil resources.[15] A 2006 royal decree mandated the country's large power plants, which were earlier planned to run on gas, would be fired by subsidized crude oil in view of declining gas production and the requirement of gas as feedstock in the petrochemical sector.

In the face of poorly developed public transportation system, fuel for private and commercial vehicles, is another large consumption arena of Saudi oil. An added demand driver for oil is Saudi Arabia's young and a fast growing population[16] with explicably diverse energy consumption patterns. With a massive construction initiative to build residential units underway to cope with the housing shortage,[17]per capita consumption of electricity in the residential sector is bound to escalate in the near future. Further in the absence of large sources and facing one of the highest per capita consumption of freshwater in the world (the country consumes 250 litres per capita each day),[18] Saudi Arabia extracts potable water from desalination plants that are huge consumers of electricity. Desalination plants typically use distillation and reverse osmosis to purify water, both energy-intensive methods. As the world's largest producer of desalinated water, the kingdom pumps out more than 3 million cubic meters of potable water each day, using 1.5 million barrels of oil at its plants.[19]

To provide a substitute for oil, Saudi Aramco is trying to notch up the natural gas production capacity, targeting both conventional and unconventional gas resources (shale), and change its power production facilities from oil resources toward natural gas.[20] Production increases of associated gas remain linked to an increase in oil production, which in turn is linked to international demand and Organisation of Petroleum Exporting Countries' (OPEC) production quota. Such vagaries of oil production have a bearing on production expansions for associated gas. Natural gas production remains limited, as soaring costs of production – exploration, processing, and distribution – at controlled low domestic prices have squeezed supply. Moreover, efforts to find new sources of conventional gas resources in the Rub al-khali – the 'Empty Quarter' region in Saudi Arabia's southeast – with foreign joint ventures, have failed to find commercially viable deposits beneath the vast expanse of sand dunes.

Consequences of Domestic Consumption

Saudi Arabia urgently needs to reconcile its domestic energy requirements with export imperatives. Several studies warn that the kingdom could be on a trajectory of declining exports in the future, if internal demand remained unchecked.[21] A threat to oil export capacity generated by domestic demand could compromise the kingdom's role as swing producer in the global oil markets that has a stabilizing impact both on energy availability and prices. In 2011-12, Saudi Arabia produced more oil than usual to counterbalance the decline in Iranian oil exports that resulted from a Western embargo in the wake of the nuclear standoff. For instance, the country produced 11.2 mb/d of oil and natural-gas-liquids in 2011, 13 percent of the world's supply and more than any other nation.[22] In August 2012 alone, Saudi Arabia produced 9.9 mb/d, the highest in the world, and more than 10 percent of global demand.[23]

A Citigroup report – "Saudi Petrochemicals: The End of the Magic Porridge Pot?" – authored by Heidy Rahman, an analyst at the bank, similarly goes further to say that Riyadh could become a net oil importer by 2030, to keep its young and growing population energy-satiated, if no changes were made to current consumption trends.[24] A reduction in crude oil burn is critical owing to the high opportunity cost associated with it. In the past several years, the price of oil has been rising; less oil in the market entails less revenue. With crude oil for electricity production is priced at $4 a barrel, the export revenue lost is more than $100 a barrel. The subsidy for power production, moreover, is just one of many fuel subsidies that the Saudis enjoy. In general, they pay $5 to $15 a barrel for their fuel from Aramco.[25] "As a result of its subsidies we calculate 'lost' oil and gas revenues to Saudi Arabia in 2011 to be over $80 billion. This could rise to $400 billion by 2035," writes Rehman.[26]

A drop in oil export capacity, which has repercussions on Saudi Arabia's revenue earnings, is a prognosis of great concern, especially as the country increases its social spending by several billion dollars, in an attempt to fend off social tension or a popular uprising like the Arab revolutions around it. The kingdom is committed to spending well over $100 billion on unemployment aids, housing construction, educational grants and health aids on its population. Regional turmoil has bestowed the Saudi government a solid rationale to increase production and curb domestic consumption.[27]

Policy Responses of the Kingdom

With oil revenues accounting for an enormous chunk of its revenues, Saudi Arabia is very cognizant of the fact that it needs to address potential fallout of unrestrained consumption. The Saudi response has been two pronged, on the demand and supply side, both with limited success so far. On the demand side, reduction of subsidy levels and initiating targeted subsidy plan of direct cash assistance to poor sections of the population is on cards. In addition, price reform is also on the agenda and the concept of 'lifeline' electricity pricing is being discussed. Saudi officials are increasingly bringing into public domain the point of subsidized and unsustainable domestic oil consumption, particularly in electricity generation.[28] However, the use of subsidized fuel as feedstock in petrochemical industry and desalination plants remains unaddressed. Beyond declaration of intent to manage the demand growth, no substantial policy measures have been initiated so far.

This is because subsidy is not just a "share out" of the benefits of energy exports by large fossil fuel exporters,[29] but also a political tool to buy acquiescence of the people in the absence of a participatory political order. The authoritarian Saudi state distributes 'rent' in the form of subsidies to perpetuate itself. Therefore, subsidies constitute the basis of the social contract in Saudi society: 'no representation without taxation,' an inversion of the classic phrase 'no taxation without representation'. 'This model of a social contract whereby key sectors of society forgo their political rights, in exchange, for access to wealth has largely held up over the past four decades,' explains Christian Koch.[30] Therefore, Saudi Arabia has resisted international calls to remove energy subsidies and has no concrete plans to raise energy prices for domestic consumers. In the context of growing demands for popular political participation in the Arab world, to which the kingdom has not remained immune, 'the leadership's impulse is to stress increases in benefits for the people rather than reductions – that is, the promise of new forms of power to meet demand rather than any measures to reduce that demand.'[31]

Hence, the emphasis is now on managing the supply side. Saudi Arabia, accordingly, is attempting to initiate projects in non-conventional energy sources and nuclear power. As part of its national energy strategy, the KSA plans to install 24 GW of renewable capacity by 2020, and 50 GW by 2032, with 16 GW solar PV, 25 GW solar CSP, and 4 GW from geothermal and waste, and 9 GW of wind capacity. About 18 GW of nuclear capacity by 2032 of total 123 GW is also projected. Despite these broad projections, hydrocarbons would still supplement the non-conventional sources by 60.5 GW – about half the capacity in 2032.[32] Non-conventional or renewable energy has its own downsides. Woertz maintains that these new sources of energy supply 'which invariably will be more expensive to produce,' which, 'in turn will potentially increase the subsidy burden of the country if consumer prices should be maintained at current low levels.'[33]

The kingdom's foray into nuclear power is very extensive, envisaging the construction of 16 nuclear power reactors by 2030, at a cost of more than $80 billion. The first two are planned to be on line by 2022 and then two more per year to 2030, and these would satisfy about 20 percent of the country's electricity demand. An April 2013 timeline showed nuclear construction starting in 2016, although experts cast serious doubt on the nuclear plans and the kingdom's proposed nuclear power launch of 2020 given cost overruns or high capital costs, and, more importantly, the high risk involved in keeping a nuclear power plant safe in high desert temperature, especially when there is no precedent of successful execution in such conditions.[34] In view of cost uncertainties and long construction time (global average is 8 years), even Mohammad El Baradei, then-director general of the IAEA was sceptical. In a 2007 visit to Saudi Arabia, El Baradei anticipated that "carrying it [nuclear energy plan] might take up to 15 years."[35] It is not clear if Saudi Arabia will develop uranium enrichment but the lack of indigenous nuclear technical expertise as well as sources of uranium at home indicate that any such move will be fraught with several impediments.

Saudi Arabia's Civilian Nuclear Development

A growing number of countries across the Arab world are seeking to embark on civilian nuclear energy programmes. Matthew Fuhrmann argues that this phenomenon is a post-2006 oil price hike development. He adds that nuclear supplier countries agreed to assist the Arab countries to 'enhance their energy security by swapping technology for oil.'[36] Nuclear power appears to have moved rapidly up the Saudi agenda, spurred by the UAE's contracts for four nuclear power stations and Iran's advances in this field. According to Marks Hibbs, Saudi Arabia believes its future economic security depends on diversifying its energy system away from fossil fuels. While Saudi policymakers have been interested in nuclear energy for the past three decades, it is only recently that the interest has gained political traction. Leaders at the top level of decision-making are convinced that the country would face a serious political and economic crisis in not-so-distant future, if it did not diversify its energy usage from total dependence on fossil fuels, namely, oil and gas.[37]

No country in the Middle East has a full-fledged plant generating nuclear power. After two decades of construction, Iran is the only country to have built a power reactor, which has yet to become functional. Saudi Arabia is one of the many countries in the region that has announced plans to develop a civilian nuclear programme. In 2009, UAE awarded contracts worth $20 billion to Korean firms for building four reactors by 2017, replicating a similar imperative as Saudi Arabia, of the need to diversify energy usage away from hydrocarbons. Other Persian Gulf countries – Oman, Kuwait, and Qatar – have also expressed interest in nuclear energy.[38] However, many anticipate if Saudi Arabia succeeds in setting up nuclear power reactors on its territory, these smaller states may instead choose to establish grid connections with Saudi Arabia and buy power.[39]

In 2006, the GCC – of which KSA is the largest member – commissioned an assessment study on the peaceful uses of nuclear energy, in addition to investigating the feasibility of developing a joint nuclear energy programme with the GCC. Initial efforts at the latter failed to materialise, and in August 2009, Saudi Arabia announced that it was considering a separate nuclear power programme. About a year later, in April 2010, King Abdullah bin Abdulaziz's royal decree – establishing KACARE – identified the development of nuclear energy as essential 'to meet the Kingdom's growing requirements for energy to generate electricity, produce desalinated water and reduce reliance on depleting hydrocarbon resources.'[40] The stated goal is to wean away the Saudis from their dependence solely to an energy mix of atomic and renewable energy," according to Muhammad Garwan, the atomic energy team leader of the KACARE [41]. The charge of the Kingdom's extant Nuclear Energy Research Centre was also transferred to KACARE. In July 2010, the *World Nuclear News* identified the task of KACARE as follows:

> The Riyadh-based 'city' will be tasked with promoting research, making future deals, and overseeing activities related to the use of atomic energy, according to an announcement by the Saudi Arabian government. It will also be the competent agency in charge of fulfilling national requirements in reference to existing and future treaties on nuclear and renewable energy signed by the kingdom, as well as being responsible for supervising works related to nuclear energy and radioactive waste projects. The initiative will be primarily funded through allocations from the state budget.[42]

Tasks Undertaken by KACARE

According to the Forum on Energy Report[43], the KACARE has undertaken the following tasks:

- In June 2010 KACARE hired the Finland- and Switzerland-based Poyry consultancy firm "to help prepare a draft of the national vision and high-level strategy in the area of nuclear and renewable energy applications for Saudi Arabia."
- In November 2011, it assigned Worley-Parsons to conduct site surveys to identify potential sites for the development of power plants. The process is on-going but as of May 2014, three sites were short-listed on the bases of three primary criteria: proximity to water sources (required as coolant in reactors), position of the KSA's electricity grid, and their proximity to electricity intensive consumers, such as desalinization plants. The identified locations are Jubail on the Gulf coast and Rabuk and Jizan on the Red Sea.

- In 2104, KACARE set up an independent regulator – the Saudi Arabian Atomic Regulatory Authority (SAARA) – to oversee the Kingdom's civil atomic energy programme and partner with other foreign regulators.
- As of May 2014, KACARE has not yet selected any particular company or national partner to build or operate any of its nuclear power reactors. However, it has been active in signing international agreements to facilitate the bidding and construction process.
- KACARE has also set up a nuclear holding company to serve as a private sector arm of the kingdom, which could sign joint-venture agreements with foreign partners to design and operate the nuclear sites and research reactors.

The KSA is likely to opt for a variety of different nuclear plants based on diversified technology – in order to avoid over-stretching one reactor builder and allow the kingdom to sign politically potent, long-term contracts with several of its biggest trading partners – and is keen only on Generation 3 and 3+ advanced reactors, which have already licensed, built and operated safely. Saudi Arabia plans to call for preliminary bids for its first two nuclear reactors in 2014, inaugurating the region's biggest atomic energy programme.[44] An April 2013 timeline showed nuclear construction starting in 2016.

Nuclear Cooperation and Potential Nuclear Partners

The KSA has nuclear cooperation agreements with several countries including France, China, Argentina, and South Korea. KACARE is currently negotiating agreement with the Russian Federation, the Czech Republic and the United States.[45] The latter half of 2013, saw reactor vendors preparing to pursue potential contracts in Saudi Arabia's fledgling nuclear power programme, with the announcement of two separate memoranda of understanding (MoU) both involving the nuclear operational expertise of Exelon.[46] In September, GE Hitachi Nuclear Energy (GEH) and Toshiba/ Westinghouse signed contracts with the US-based nuclear power operators, Exelon Nuclear Partners (ENP) – a division of Exelon Generation[47] – to pursue reactor construction bids with the KACARE.

A nuclear cooperation agreement was signed between Saudi Arabia and France in early 2011, allowing the two countries to cooperate in the fields of production, use and transfer of knowledge for peaceful uses of nuclear energy. The agreement opened the way for French nuclear vendors to bid on billions of dollars' worth of nuclear construction and operations project. France's Areva and EDF (Electricite de France) signed a series of agreement with Saudi companies and universities, aimed at developing Saudi Arabia's supply chain and workforce to support its nuclear energy programme, during a visit to Riyadh by French president Francois Hollande in December 2013:[48]

- The French companies signed MoUs with five Saudi manufacturers – Zamil Steel, Bahra Cables, Riyadh Cables, Saudi Pumps and Descon Olayan – with the aim to developing the industrial and technical skills of local companies to form a domestic supply chain.
- Areva and EDF also signed agreements with the King Saud University in Riyadh, Prince Mohammed bin Fahd University in Al-Khobar, and Dar Al Hekma College and Effat University, both in Jeddah, that are intended to contribute to the development of Saudi Arabia's indigenous nuclear expertise.
- EDF signed an agreement with Saudi Arabia's Global Energy Holding Company (GEHC) for the creation of a joint venture whose first task will be to carry out feasibility studies for a European pressurized Reactor (EPR), a generation3+ reactor, in the country. GEHC was established in 2011 to invest in the development of energy-related businesses.

Areva and EDF have also signed a cooperation agreement with Saudi Arabia's National Institute of Technology (NIT) located in Bahra, near Jeddah, with the aim of developing a pool of technical nuclear skills in the country. The two companies have launched a programme to train Saudi technicians in various nuclear specialties with an aim to provide an understanding of the safety and quality requirements specific to the nuclear industry. They have also taken representatives from several Saudi higher education establishments to their industrial sites in France to show them fuel cycle activities, operation of nuclear power plants and reactor construction. Policymakers at the highest level of decision-making have encouraged building a pool of indigenous technical talent of manufacturers and engineers.

China National Nuclear Corporation (CNNC) has also expressed interest in Saudi Arabia's nuclear power programme. During a visit to King Abdulaziz City of Science and Technology (KACST) – home to both Saudi Arabia's national science agency and its national laboratories – CNNC general manager Qian Zhimin discussed cooperation in the field of nuclear science and technology. Zhimin also signed a memorandum of understanding on cooperation with the Saudi Technology Development and Investment Company (TAQNIA). A January 2012 agreement with China relates to nuclear plant development and maintenance, research reactors, and the manufacturing and supply of fabricated nuclear fuel. China appears to be preparing to play a larger role in the global nuclear industry. In recent years it has been active in acquiring uranium assets abroad as well as obtaining advanced Western nuclear technology, which it hopes to begin exporting during the coming decades.[49]

In mid-2011, Saudi Arabia and Argentina signed a nuclear cooperation agreement focused on the development of smaller nuclear reactors. The agreement would pave the way for Argentina's Atomic Energy Commission and private technology firm INVAP to begin the construction of a simplified pressurized water reactor (CAREM) aimed at small-scale electricity generation and water desalination projects – along the Saudi coastline – which are both urgent needs for the oil-rich Kingdom. INVAP has previously built research reactors in Algeria and Egypt.[50]

A November 2011 agreement with South Korea called for cooperation in nuclear research and development (R&D), including building nuclear power plants and research reactors, as well as training, safety and waste management. In June 2013 Korea Electric Power Corporation (KEPCO) offered support for the localization of nuclear technology, along with joint R&D of nuclear technologies, if Saudi Arabia purchases South Korean reactors. In December 2009, the UAE awarded a South Korean consortium a contract to build four nuclear power plants worth $20.4 billion.[51]

In May 2008, US President George Bush promised King Abdullah that the United States would 'assist the Kingdom of Saudi Arabia to develop civilian nuclear power for use in medicine, industry and power generation.' The White House added, 'The US and Saudi Arabia will sign a Memorandum of Understanding in the area of peaceful civil nuclear energy cooperation. This agreement will pave the way for Saudi Arabia's access to safe, reliable fuel sources for energy reactors and demonstrate Saudi leadership as a positive non-proliferation model for the region.'[52] To begin any nuclear project collaboration with the US companies, the Saudi and the US governments must negotiate a so-called 123 Agreement. Section 123 of the 1954 US Atomic Energy Act requires a specific agreement for transfer of nuclear energy-related materials, equipment or components from US to another country.

So far, the parties have reached an understanding on every other issue except for the uranium enrichment and plutonium reprocessing (ENR) provision. These technologies, which can provide both fuel for nuclear power and fissile material for nuclear weapons, is presently "being studied" by KACARE. The US insistence on the so-called 'Gold Standard' provision – proscribing the agreeing party from ENR[53] – has become a major obstacle to the proposed 123 agreement as Riyadh's ENR rights under the NPT have become a matter of principle linked to the kingdom's sovereignty.[54]

Ever since the UAE chose to voluntarily forego its ENR rights in a 123 agreement signed with Washington in 2009,[55] the Congress wary of proliferation in one of the most unstable regions in the world has insisted that the Gold Standard must be upheld as a model for peaceful technology cooperation with other countries.[56] The matter has become complicated with November 2013 interim agreement between Iran and the world powers that has not only granted Tehran de-facto clearance to enrich some of its uranium, but also brought on-going US-Saudi 123 negotiations to a halt. It is unlikely that Saudi Arabia will accept recognition of Iran's enrichment programme, while having to compromise its own ENR 'rights'. US nuclear vendors are concerned that if Saudi Arabia does not ink a 123 agreement with the United States in time for the tender, they could miss out on the world's biggest nuclear building markets for the next two decades.

Nuclear Industry and Localization

The proposed expanse of the Saudi nuclear programme indicates that the goal of this project is not only to build nuclear power plants, but also build a nuclear industry. Nuclear Engineering International (NEI) reports that at least 60 percent of the nuclear spend would have some kind of localization requirements; only 40 percent would go to 'non-localized opportunities.' Localization would involve technology transfer, licensing agreements, blueprint transfers, joint ventures, development of local manufacturing plants, building EPC (engineering, procurement, construction) capabilities, focussing on operations and maintenance, and development of plans to set up personnel training programmes and centres.

NEI also reports that three methods of localization are planned: from the public sector, private sector and public-private partnership. On the basis of a conference of key executives involved in the development plan for a civil nuclear industry, NIE also reports that the localization process is foreseen in three phases.

> The first, starting from 2014-2023, will cover small-scale non-nuclear grade components such as heat exchangers and switchgears; construction (raw materials and labour); non-nuclear grade pipes and pumps. The second wave of localization, lasting until 2032, would include nuclear-grade pipes, all types of valves, nuclear-grade pumps, and control rod drive mechanisms. The third wave, lasting until 2045, would include skilled construction labour and construction equipment, and engineering jobs with nuclear expertise. The end goal, at some point after 2045, would be for self-reliance in nuclear turbines, reactor design, heavy-forged nuclear components and emergency diesel generators.[57]

'We see unique benefits of diversifying acquired technologies in terms of job creation, value chain localization, and knowledge transfer,' says Waleed Abulfaraj, Vice President of KACARE.[58] Muhammad Garwan, the atomic energy team leader of KACARE lists two advantages that would accrue with localization: 'it will enhance economic viability of the project and second it would make jobs available for the Saudi youth.'[59]

Saudi Arabia's Nuclear Programme Controversy

The risk of proliferation cannot be ruled out completely in nuclear energy programmes. 'A civil nuclear energy programme has an implied military deterrent value because it keeps opponents guessing about whether the state also has a hidden nuclear weapons capability,'[60] says Ian Jackson. Physicist Amory Lovins and attorney Hunter Lovins point out in their book *Energy/War: Breaking the Nuclear Link*, 'Every known route to bombs involves either nuclear power or materials and technology which are available, which exist in commerce, as a direct and essential consequence of nuclear power.' Further, 'All nuclear fission technologies both use and produce fissionable materials that are or can be concentrated. Unavoidably latent in those technologies, therefore, is a potential for nuclear violence and coercion which may be exploited by governments, factions,' they write. Civilian nuclear power technology, they emphasize, provides the way to make nuclear weapons, furnishing the materiel and personnel.[61]

In November 2006, the IAEA named six Arab countries – Algeria, Egypt, Morocco and Saudi Arabia, Tunisia and the UAE – as having also shown interest in development of civilian nuclear technology. In the backdrop of the controversy over Iran's nuclear programme, many independent experts argued that the surge in interest vis-à-vis nuclear energy in the Middle East represents a reaction by the countries to 'Iran's uranium enrichment and questionable activities that could be related to a weapons program.'[62]

Some experts suggest that the drive for civil nuclear power in the region is seen as a long-term 'security hedge', particularly in the light of Iran's ambitious and ambiguous nuclear programme. It has created the perception that having nuclear energy facilities – especially the fuel cycle facilities – gives the country a sense of power and strength. In the light of concerns that US security guarantees for its Middle Eastern allies might not remain forever, there is a sudden rush for developing nuclear programmes in the region. 'It's a longterm hedging option,' says Hibbs, 'particularly in light of concerns that U.S. security guarantees will become weaker.'[63]

Since the mid-1990s, unconfirmed media reports have periodically circulated about Saudi Arabia's alleged attempts to develop/acquire nuclear weapons. Since 2003, analysts believe, Saudi Arabia has maintained a veiled nuclear military nuclear strategy. The London-based *Guardian* newspaper reported in September of the same year that Saudi Arabia had embarked on a review of its security that includes acquiring nuclear weapons, apparently in the backdrop of international apprehension over Iran's alleged nuclear weapons programme. A strategy paper, supposedly considered at the highest levels of the government in Riyadh, specifies three options for strengthening national security: 1) to acquire own nuclear deterrent capability; 2) to maintain or enter into an alliance with an existing nuclear power that would offer protection; 3) and to try to reach a regional agreement on having a nuclear-free Middle East. Even as the story was denied by the Saudi political establishment with usual reiterations of making the Middle East a zone free of Weapons of Mass Destruction (WMD), Riyadh's actions over the years have caused increasing US concern about possible Saudi proliferation.[64]

Security Perception and Indigenous Nuclearisation: Aspiring for Deterrence?

Some analysts talk about security concern as the justification that would drive Saudi Arabia to seek nuclear capability or procure nuclear weapons. Yet others talk of 'national pride, popular domestic enthusiasm for advanced technology and higher geopolitical status'[65] that drives the acquisition of nuclear technology in Gulf states including Saudi Arabia. The two factors dovetail, such that public opinion in Saudi Arabia is in favour of developing nuclear weapons.

What makes Saudi Arabia vulnerable to security threats? Several geostrategic factors make Saudi Arabia's external security environment inherently vulnerable. The geographical expanse of the country coupled with thin population as compared to its rivals in the Persian Gulf – Iran and Iraq –handicaps any attempt to attain conventional military parity with them. In addition, the competence of its armed forces is untried and untested. The kingdom also has no 'strategic heartland' into which its defence forces retreat and regroup. Saudi Arabia is the most vulnerable along the Gulf coast, where its major oil fields and crucial desalinization plants are located, leaving them exposed to potential foreign attacks.[66]

Security analyst Richard Russell points out that the Saudis have been particularly fearful of foreign invasion since Iraq's incursion into Kuwait in 1990. It exposed the country's vulnerability to ground attacks and showed the dubious capability of its armed forces. What was most excruciatingly revealed for the kingdom in the wake of the Iraq-Kuwait conflict was its complete dependence on the United States for security. Would the American respond with the same vigour and conviction each time Saudi Arabia faced a major security threat?

Russell argues that although the Saudis were saved by the United States from being directly impacted by the Gulf War, they 'would be loath to rely on American benevolence or a confluence of interest in a future conflict.' Even before the Iranian nuclear ambition started to consume Saudi Arabia, Russell had pointed out, 'it would be imprudent, to say the least, for Riyadh to make the cornerstone of [its] national-security posture out of an assumption that the United States would come to the kingdom's defence - under any and all circumstances.' From Riyadh's perspective, Russell argued, 'the acquisition of nuclear weapons and secure delivery systems would appear logical and even necessary.'[67]

While Saudi Arabia has purchased advanced mobile conventional weaponry to bolster air and ground forces and use them in case of aggression, a nuclear deterrent option is the ultimate security requirement of a state lacking a spectrum of capabilities to draw on should the former fail. In the 1960s and 1970s, Saudi fears were heightened with the news of Israeli nuclear arsenal; with the turn of the century, Iran's nuclear programme has become a source of trepidation for Saudi Arabia. Saudi Arabia's fear of encirclement by adversaries intensified with the US invasion of Iraq (2003) that removed the buffer against Iran, its long-standing rival.

Saudi Arabia's secret purchase of CSS-2 (Dongfeng3) missiles from China in the mid 1980s gives us some clue of Saudi insecurity vis-à-vis its neighbours. Dongfeng3 (DF-3) is road mobile, liquid fuelled, medium-range ballistic missile (MRBM). With a range of 2,500 km, the DF-3 has extensive regional reach. The DF-3 is a highly inaccurate missile; when equipped with a conventional warhead, it would therefore not be effective against discrete military or tactical targets. Although the DF-3 missiles were originally designed by China to carry a nuclear payload, they were apparently modified to deliver conventional high explosives for Saudi Arabia. The country has never tested the DF-3, which would be critical for ensuring its reliability and training missile forces. It is reasonable to believe that the DF-3 is not currently an active component of Saudi defence strategy.[68]

As DF-3 missiles have reached the end of their service life, China has replaced them with nuclear and conventionally armed solid-fuelled CSS-5 (DF-21) ballistic missiles. It is hardly surprising, then, that Saudi Arabia also sought to do the same thing. Reportedly, the sale of DF-21 missiles to the kingdom took shape in mid-2007 with the CIA backing, which apparently examined the weapons in their packaging, and determined that they were not intended to hold nuclear payloads. Analysts differ in their assessments of what advancement in capabilities the midrange Df-21 would offer over Df-3 missiles. Some say the weapons are more accurate and can be launched on shorter notice. According to another expert, the impact of the reported sale would be minimal if the missiles are not tipped with nuclear warheads.[69] Theoretically, it is possible to fit these missiles with nuclear warheads acquired from China, Pakistan, or North Korea but as Lippmann observes, "Any attempt to do so, however, would present immense technical and political difficulties – so much so that Saudi Arabia might emerge even less secure."[70]

Apprehensions that Saudi Arabia might develop an indigenous military nuclear capability, if it establishes a civilian nuclear programme, do not have solid bases. Since the beginning of the nuclear age, the nuclear weapon powers in the West, especially the US, 'has regularly feared that the world was approaching a nuclear tipping point. The fact that a nuclear cascade has failed to materialize has done little to dampen these fears.'[71] Despite apprehensions that they can be acquired rather easily, nuclear weapons are not commonplace. Several technical barriers hinder the entry of a state to the 'weapons club' in addition to the difficulty of hiding a military programme completely from foreign intelligence observers.[72]A country must have natural resources and technological capability at command to master the several facets of the nuclear fuel cycle – mining of uranium, conversion, enrichment and reprocessing –in order to develop a nuclear weapons programme. Saudi Arabia does not have uranium deposits on its territory, so it lacks experience in uranium mining. Since the country does not operate nuclear power facilities yet, its scientists also lack experience in nuclear fuel conversion and fabrication, enriching uranium for reactor fuel, and operating nuclear reactors.[73] There is no evidence either that they are planning to set up centrifuges for indigenous uranium enrichment.

Moreover, As Ian Jackson says:

> ... the proliferation of military nuclear weapons does not automatically follow the acquisition of nuclear energy technology. Many countries possess advanced nuclear skills but have not deployed nuclear weapons systems. Any country with significant nuclear technology should probably be regarded as a latent nuclear weapon power. Yet only one-third of the 30 countries that possess civil nuclear

> energy programmes have weaponized them ... it seems proliferation is a choice, not a certain outcome of a state nuclear energy programme.[74]

It is highly unlikely that Saudi Arabia would consider proliferation as long as it shares a strong strategic-security relationship with the United States. If it did so, the Kingdom would lose US security umbrella, and assistance on civilian nuclear power, mentioned in the MoU signed with Washington in 2008. Saudi Arabia would also lose arms upgrades and sales from Washington on which the country's military superiority is based. Moreover, as Lippman argues, nuclear proliferation would involve confrontation with the West and defiance of the global proliferation regime will desist from doing by the very nature of its functioning:

> ... the Saudis' weapons of choice are cash and diplomacy. It is difficult to imagine the princes of the House of Saud deliberately positioning themselves as global outliers and inviting reprisals from countries capable of inflicting serious political and economic damage on them. With hundreds of billions of dollars of private Saudi capital and government funds invested in the United States, the Saudis would be ill-advised to risk an asset free.[75]

The high economic cost of acquisition of nuclear weapons is also repugnant to Saudi Arabia. Kate Amlin of the Monterey Institute of International Studies (California, US) believes the Saudis 'do not want to incur the political and economic backlash that would result from their development of a nuclear arsenal at a time when they are trying to integrate more into the international economy.'[76] In the context of major industrialization efforts and integration with the global economy, 'it's not rational that they would jeopardise this in favour of a pre-emptive strike against the theoretical possibility of a nuclear-armed Iran.'[77]

However, the Saudi officials are extremely worried that the growing political influence of Shia populations led by Iran in countries surrounding their kingdom – Iraq, Bahrain, Yemen and Syria – will become an existential threat to their own regime's survival. Saudi leaders perceive 'Iran's policies as part of an expansionist, sectarian agenda aimed at empowering Shia Muslims in the region at the expense of Sunnis,' a CRS study concludes.[78] Iran's nuclear programme is viewed within Saudi Arabia as physical proof of the growing Shia threat in the region, which heightens fears of 'unbridled Iranian support for an active "Shia Crescent" from the Levant through the Eastern Province of Saudi Arabia to Bahrain.'[79] Thus, the Iranian nuclear issue is a particularly important point of tension between Saudi Arabia and Iran.

The statements emanating from the Saudi political establishment have strengthened the notion of 'Iran factor' in the 'sudden rush' of Arab countries towards a nuclear programme. King Abdullah of Saudi Arabia warned visiting US special envoy to the Middle East Dennis Ross in 2009 that if Iran crossed the threshold, 'we will get nuclear weapons.'[80] A similar sentiment was echoed by Prince Turki al-Faisal, a senior Saudi diplomat and former head of Saudi Arabia's General Intelligence Directorate, in a meeting with NATO officials at Royal Air Force Molesworth, UK. He stated that 'if Iran develops a nuclear weapon, that will be unacceptable to us and we will have to follow suit.'[81]

Saudi Arabia's fear is not so much that Iran would use nuclear weapons for an actual physical attack but that such weapons would embolden Tehran to 'pursue its own, more hegemonic foreign policy in the region,'[82] so as to change the balance of power in the Middle East to the disadvantage of Riyadh. Abdulaziz Sager, head of the Geneva-based Gulf Research Centre, argues that nuclear weapons would turn Iran into

> A hegemonic power over the [Persian Gulf] states of the region, through its control of Iraq, its holding fast to the continued occupation of the UAE's islands, and its intervention in the domestic affairs of countries in the region through the agitated Shiite groups in these countries, which could push the GCC, namely Saudi Arabia, to seek, in turn, the acquisition of a nuclear weapon to confront Iran.[83]

Lewis Dunn aptly summarized the concerns of Gulf countries including Saudi Arabia in a paper for the US Naval Postgraduate School (2009), 'Among virtually all of Iran's neighbours, there is considerable concern that once in possession of nuclear weapons, the Iranian leadership could intensify its interference in those neighbours' domestic politics, step up support for extremist groups, and use the shadow of nuclear weapons to shape regional relations in a manner more favourable to it.'[84] 'The issue for the Saudis is not merely Iran's putative nuclear capability. A deal on Iran's nuclear program would legitimize the regime's regional influence in a way that has not occurred in decades, thereby serving its hegemonic objectives. The deeper threat or fear is that Iran's ultimate target is leadership of Mecca, the cradle of Islam,' says Mai Yamani.[85]

The Kingdom has expressed strong reservations about efforts by P5+1 (led by Washington) to negotiate a deal with Iran on its controversial nuclear weapons programme. In the wake of the interim Geneva deal of November 2013, Saudi Arabia expressed the fear that Iran could be left with a break out capacity to make the fuel for a nuclear weapon at a time of their choosing. While the Saudis recognize there is no easy military response to Iran's nuclear programme and no military strike can eliminate the country's advancement in nuclear knowledge, they are irked with Obama's decision to ease sanctions, instead of tightening them further and possibly causing the regime to collapse.

Saudi Arabia, views the 'developments in Iran very negatively,' says David Albright, the president of the Institute for Science and International Security and a former weapons inspector. 'The Saudis are thinking through how do you create a deterrent through capability,' he adds.[86] In February 2012, a senior Saudi official told the London-based *Times*:

> There is no intention currently to pursue a unilateral military nuclear programme but the dynamics will change immediately with the Iranians develop their *Own nuclear capability* ... Politically, it would be completely unacceptable to have Iran with nuclear capability and not the kingdom.[87]

Many fear that petro-dollar power of Saudi Arabia would easily allow it to buy talent and training for a weapons programme more easily, when an indigenous Saudi program is in early stages. A RAND analysis sums up Saudi fear how, 'The advent of nuclear-armed Iran would likely be perceived as an existential threat to Riyadh, possibly pushing Saudi Arabia to acquire its own countervailing deterrent.' As far as Saudi Arabia is concerned, 'any plan to develop nuclear power will inevitably raise concerns about the parallel issue of acquiring nuclear weapons – largely because of perceived threat from Iran.'[88]

Fear of a 'Saudi bomb' has resurrected in the wake of Riyadh's differences with Washington over Iran and Syria. The US security umbrella has to date provided the Saudis with unprecedented level of protection, which they believe may be weakening over several strategic differences over events in the region. Making Riyadh's insecurity worse is the declining importance of Saudi oil for America's energy needs. New energy supplies – particularly shale oil and gas in the US and Australia – have diminished the United States' direct dependence on the Kingdom, although its European allies still depend in a major way on the Middle Eastern energy supply.

A major deterioration in US-Saudi relations leading to withdrawal of Washington's security guarantees could compel Riyadh to take the nuclear weapons option more seriously. In such a scenario, Saudi policy makers believe, a nuclear-weapons-capable Iran would be a grave strategic threat to the kingdom as well as the region. It would undermine Saudi regional influence, and threaten the domestic stability of the Gulf countries through incitement of their Shia population. Nonetheless, Saudi domestic arena would still be threatened not the least by Iranian nuclear capability. Once the relations between Iran and the West are normalized in the event of a nuclear agreement in 2014, a free flow of Iranian oil may depress oil prices, impacting Saudi Arabia directly, as the government depends almost exclusively on oil export revenue for its budgetary spending. Dissatisfied with Washington's response to Syria and Iran, and beset with a deep sense of insecurity over its seeming decline in regional influence, Saudi Arabia has moved to illustrate this by carrying out its largest-ever military exercise, with the participation of Pakistan, a close ally and plausible nuclear guarantor.[89]

Dealings with other Nuclear Powers: The 'Pakistan Factor'

Some experts worry that Saudi Arabia could overcome its domestic resource inadequacies by acquiring nuclear weapons from outside assistance - perhaps from Pakistan or China. Saudi Arabia is known to have funded Pakistan's nuclear weapons (for the production of the so-called 'Islamic Bomb' or the so-called 'Sunni Bomb'), by directly financing the research of A.Q. Khan, the father of Pakistan's missile programmes. In his book, *Eating the Grass*, a semi-official history of Pakistan's nuclear programme, Major General Feroz Hassan Khan acknowledges that Saudi Arabia provided generous financial support to Pakistan that enabled the nuclear program to continue.[90] It is often reported that the kingdom could mobilize those weapons from Pakistan, if needed, a country that has sent military and security assistance to Saudi Arabia from time to time.[91] Although there is no official confirmation of such recurring reports by either the Saudi or Pakistani sides and it is not clear how such a deal will unfold, experts consider that it could work in three ways: acquisition of atomic warheads, pledge of nuclear cover, and/or deterrent guarantee by actual deployment of Pakistan's nuclear weapons.[92]

BBC's Mark Urban reported in November 2013 on *Newsnight* (a weekday BBC Television current affairs programme), Saudi Arabia believes it could obtain bombs 'at will' from Pakistan, quoting intelligence that 'nuclear weapons made in Pakistan on behalf of Saudi Arabia are now sitting ready for delivery.' Security analysts and US diplomats hold that Saudi Arabia in effect already has nuclear weapons in the nuclear arsenal of its long time friend, Pakistan. For instance, Gary Samore, who until March 2013, was President Barack Obama's counter-proliferation adviser, spoke in the same programme thus: 'I do think that the Saudis believe that they have some understanding with Pakistan that, in extremis, they would have claim to acquire nuclear weapons from Pakistan.'[93] Samore qualifies elsewhere that while the Saudis have delivered huge amount of money to Pakistan, and some went to support the nuclear weapons programme, there is improbability of an explicit *quid pro quo* deal. 'What [is] more likely would be that Pakistan would [again] station troops on Saudi soil, and those could include nuclear-armed forces.'[94]

Despite the reports, Pakistan has compelling reasons not to transfer such weapons to Saudi Arabia. A transfer of weapons to Saudi Arabia would evoke stringent measure by the United States, such as disbursal of funds from the International Monetary Fund (IMF), which helps keep Pakistan's solvency. Pakistan is keen to maintain good relations with Washington in order to continue receiving $1.5 billion of annual civilian aid under the Enhanced Partnership with Pakistan Act of 2009. Any proliferation activity would lead to the discontinuation of this assistance as well as other military partnerships between Washington and Islamabad. Moreover, Pakistan's nuclear programme is directed as a deterrent against India, which would be jeopardized if its nuclear arsenal were to move out of the country.

Iran would be even more infuriated by the move, cancelling all plans to export oil and gas to energy-starved Pakistan. Iran is also capable of stirring trouble in Pakistan's troubled Baluchistan province as well as Afghanistan, especially with the help of Islamabad's archenemy India.[95] Further such a move would provide immense legitimacy to Iran's nuclear aspirations and intensify its urge to acquire nuclear weapons, squeezing Pakistan between two hostile nuclear-armed states. In addition, Saudi nuclear acquisition would infuriate and hasten an Israeli preventive strike, especially when the weapon is being prepared. With the nuclear *bazaar* of Pakistani scientist A.Q. Khan dismantled, the clandestine transfer has become a distant possibility.

At a theoretical level, the Saudis could pursue a nuclear option with China, but in the current security this also seems unlikely. Despite the Chinese missile transfer to Saudi Arabia, Beijing and Riyadh never have had close military ties because of Washington presence in the Middle East and consequent security assurances to all the Gulf countries, in addition to providing the Saudis with high technology defence equipment. Moreover, it is even more unlikely that China would jeopardize its several relations with Washington to help Saudi Arabia have nuclear weapons. China is also a member of the NPT and obliged not to assist a non-nuclear state in acquiring nuclear weapons. Moreover, under the US-led Nuclear Prevention Proliferation Act (NPPA) of 1994, Beijing would face the revocation of the US nuclear cooperation agreement it worked so hard to secure as well as possible imposition of economic sanctions.

Given the lack of transparency in the Saudi decision-making process, it is impossible to know whether there have been discussions on nuclearisation at the highest level in Riyadh. In the current scenario, Saudi Arabia does not face any current external threats. Given that there are written and unwritten agreement between Saudi Arabia and the United States that would provide protection to the country in case of threats to state security, the possible benefits of acquiring nuclear weapons 'would be psychological and political rather than strategic, including potential deterrent effect against reckless neighbours.'[96]

Middle East: Nuclear Weapons free Zone (NWFZ)?

Since Iran proposed and the United Nations General Assembly (UNGA) adopted a resolution supporting the establishment of a Middle East NWFZ in 1974, Saudi Arabia, at least, officially has consistently supported this idea as an alternative way to ensure complete security. It is a party to the NPT and has signed a comprehensive safeguards agreement with the IAEA. Saudi Arabia, like Iran, has not yet signed the Additional Protocol, which allows for stricter inspections. Nor has it signed the Comprehensive Nuclear Test Ban Treaty (CTBT). After the GCC programme was announced in 2006, Prince Saud Al-Faisal, the foreign minister of Saudi Arabia, told reporters, 'We want no bombs....Our policy is to have a region free of weapons of mass destruction.'[97]

Saudi officials have targeted Israel's undeclared nuclear weapons arsenal, as the main impediment to the establishment of a Middle East free of nuclear weapons. With Iran's nuclear activities expanding since the 1990s, Saudi officials have chastised Iran for its nuclear efforts. However, they have been reluctant to confront Iran directly over its nuclear programme. While Israel still has to declare its nuclear programme and commit to arms control and reductions, on the issue of Middle East NWFZ, in the least, Saudi and Iranian views accord officially. Iran may still be open to the idea, from its initial sponsorship during the Shah's regime, because of the fact that it has already mastered the capability to run the entire fuel cycle.

In a speech delivered in April 2014 before the meetings of the Preparatory Committee for the NPT Review Conference, 2015, Ambassador Abdullah bin Yahya Al-Mo'allemi, reiterated the oft-repeated distinct positions on the Israeli and Iranian nuclear projects. While expressing concern over the intractability of the region in moving on the issues of eliminating WMD, he squarely blamed Israel for thwarting any efforts in this direction. On the Iranian nuclear issue, which has seized Saudi Arabia's security establishment for long, Mo'allemi demanded aggressive IAEA inspections to Iran's nuclear sites and clear timeframe for negotiations between the Group 5 +1 and Iran.[98]

Earlier, Turki al-Faisal in an interview to the *Associated Press* in January 2012, exhorted the international community to help establish a NWFZ in the Middle East. The idea 'deserves everybody's attention and energy,' he declared, warning of the potential for a nuclear arms race in the region if it was not implemented. He urged the five permanent UN Security Council members to guarantee a nuclear security umbrella for Middle Eastern countries that agree to a nuclear weapons-free zone and impose sanctions on countries that develop or maintain nuclear weapons. The former Saudi intelligence chief was referring not just to Iran – on whom the Security Council has already imposed sanctions for its nuclear programme – but also to Israel, the only state in the region that possesses nuclear weapons, although they refuse to confirm or deny the fact.[99]

Turki al-Faisal outlined a 'political plan for a Weapons of Mass Destruction-free Zone (WMDFZ) in the Middle East,' in a paper published by Harvard's Belfer Centre in July 2013. He said:

> No region in the world is in greater need of becoming such a zone than the conflict-cursed Middle East. This region is in the midst of a drastic transformation. One of its states (Israel) is a de facto nuclear weapon state and another one (Iran) is progressing toward becoming one. Both countries have lied about their intentions in this regard. Israel publicly stated that it would not introduce nuclear weapons into the Middle East, yet it has over 200 warheads with the required delivery systems. Iran has publicly stated that it seeks the peaceful development of nuclear power, yet it continues to raise enrichment levels in its centrifuges while increasing their number and preventing IAEA inspectors from inspecting suspected sites.[100]

In 2012 the establishment of a WMDFZ in the Middle East was expected to return to the top of the international agenda as Finland was scheduled to host an international conference on the issue, with the co-sponsors being the US, UK, Russia and the UN Secretary General. The meeting was postponed indefinitely due to 'present [Arab Spring revolutions] conditions in the Middle East and the lack of agreement among the participating states, on 'acceptable conditions'.[101] The decision to hold the conference was taken at the 2010 NPT Review Conference, following repeated entreaties by Middle Eastern states to ascribe the zone issue to the agenda of the NPT. At the Review and Extension Conference of the NPT in 1995, a resolution adopted calling for practical steps to establish weapons free zone in the Middle East, constituted a major part of the bargain that saw the NPT extended indefinitely. Since 1995, however, little progress has been made, and at the 7th NPT Review Conference in 2005, Egypt blocked any final agreement citing a lack of progress or intent on the zone question.

In October 2013, Saudi Arabia rejected the non-permanent UN Security Council seat – citing the recent impasse over the Syria conflict as well as the lack of progress on a move towards banning WMDs in the Middle East. Apart from Israel's refusal to give up its nuclear weapons and join the NPT as a non-nuclear-weapon state, Iran's nuclear enrichment programme adds further complexity to the issue, and as mentioned before Saudi Arabia has indicated several times that if Iran crosses the threshold, it will be compelled to take the nuclear route. Egypt and Syria, in reaction to Israel's nuclear weapons, have refused to ratify the two other conventions dealing with WMDs (on biological and chemical weapons). With several key states tied up in their own positions, some claim that the only viable option is to reinvigorate the process of Arms Control and Regional Stability of the early 1990s.[102]

Conclusions

The Kingdom of Saudi Arabia, a country usually associated with plenitude of hydrocarbon-based energy resources, is on the path of developing one of the most extensive nuclear power programmes in the world. Soaring domestic demand for oil has created apprehensions about the decline in the country's spare capacity that could hamper exports and destabilise the global oil market. This has prompted consideration of non-conventional energy sources to supplement internal demand that would free up more oil for export. Saudi Arabia has charted out a comprehensive non-conventional energy programme in which nuclear energy is a crucial segment. While Saudi Arabia has signed nuclear cooperation agreements with several countries that will allow them to assist the KSA in developing civilian nuclear technology, it has not concluded the 123 agreements with the United States. Failure to do this will debar the US nuclear companies from undertaking certain activities in the KSA's nuclear power project. It will also hinder other nuclear suppliers from transferring enrichment and reprocessing technologies to Saudi Arabia as well as strip the kingdom of the 'legitimacy' and 'acceptability' of its programme among the nuclear vendors. Some of the preliminary steps on ground have been taken to launch the nuclear programme but the kingdom is yet to summon bids for building nuclear reactors in view of concerns about safety and nuclear waste.

Saudi Arabia's nuclear power development plans have evoked fears and generated controversies, mainly because civilian nuclear establishment can possibly be used for weaponisation, the Iranian nuclear imbroglio being a case in point. Although enrichment and reprocessing is not envisaged at this moment in the civilian nuclear programme, they remain looming questions, especially as nuclear weapons remain a part of Saudi strategic thinking. In the foregoing its has been argued that Saudi Arabia will be inclined towards acquiring nuclear weapons or nuclear weapons deterrence cover under two dire circumstances: first, if Iran goes nuclear and, second, if special relationship between Saudi Arabia and United States deteriorates to the point of no-return. While the possibility of Iran developing nuclear weapons seems to have receded with the Geneva agreement, it seems to have ironically contributed to Saudi insecurity about the America's commitment to its security. Instead of appreciating the reduced threat of a nuclear-armed Iran – within striking distance across the Persian Gulf – Saudi Arabia, the regional Sunni Muslim power, has grown ever wary of what it perceives as a diplomatic thaw between its Shia Muslim rival and the US.

Saudi Arabia has been advocating a total dismantling of Iran's nuclear programme, preferably in US airstrikes on Iran's nuclear facilities. But as the on going diplomacy suggests, Iran would be allowed to keep its nuclear programme and would retain the right to enrich uranium. The Saudis understand that Iran by virtue of its advanced nuclear programme, retains the knowledge and capability to produce nuclear weapons – what an agreement might do is to extend the breakout period for a duration that any rush towards nuclear weapon is detected and stopped. The situation, as it exists, is repugnant to the Saudi perception of regional security, which is partially based on keeping Iran isolated and contained.

In the last few years, Saudi Arabia has become increasingly disaffected by the regional policies of its security guarantor, the United States. It viewed with trepidation the Obama administration's 'abandonment' of former Egyptian president Hosni Mubarak, America's long-standing ally and its subsequent acceptance of the Muslim Brotherhood government in the name of supporting democracy. President Barack Obama also refused to intervene militarily in Syria to help rebels fighting against President Bashar Assad's regime, who are supported by Saudi Arabia. But what has really thrown the Saudis off-balance is America's support for the interim agreement on Iran's nuclear programme, which marks a diplomatic thaw between Washington and Tehran.

With the Saudi leaders anxious of Iran's rising power in the region and beset with their own internal problems with Shias, people's demand for reforms, as well as the 'threat' of a resurgent Muslim Brotherhood, the notion that the US commitment to the region is wavering has been gaining ground. In November 2013, senior Saudi figures threatened a 'major shift' away from Washington, warning that Riyadh might 'go it alone' if Obama continued to neglect Saudi demands for coercive action on Syria and Iran. After the twin debacles of Afghanistan and Iraq, president Obama has sought to redefine US policy as 'minimalist, multilateral when possible, and unabashedly self-centred,' says Gary Sick, a senior research scholar at Columbia University's Middle East Institute. 'Obama, unlike Bush, is not willing to go in and change the face of the Middle East to suit Saudi desires, to get involved in nation building and in basic domestic issues in countries,' Sick adds.[103]

With Iran retaining the capability to build nuclear weapons, if not actual warheads, in the event of a nuclear deal, the question upmost in Saudi mind is whether the US will retain its long-term commitment in the region to which the kingdom's security is inextricably linked. Robert Hunter, a RAND analyst,[104] says that the commitment of the US is bound to diminish because of high cost of resources both in terms of men and material required to fulfil it. A comprehensive security arrangement is not possible in such a fractious region, he argues, but proposes partial steps, similar to those developed with the Soviet Union during the Cold War, in which Iran could be an interested party. The critical aspect for Saudi Arabia is whether they are perceived as a 'valuable and reliable partner' for the United States and would continue to be so in the years ahead.

Notes

1 "Saudi Arabia: Country Analysis," *Energy Information Administration (EIA)*, September 10, 2014, http://www.eia.gov/countries/cab.cfm?fips=SA; *BP (British Petroleum) Statistical Review of World Energy*, June 2014, p. 9.

2 Saudi reserves are about 266 billion barrels (bb) or 19 percent of world total oil resources; the top reserve country is Venezuela with 298 bb of oil. *BP Statistical Review of World Energy*, June 2014, p. 6; Rupert Rowling, "Venezuela Passes Saudis to Hold World's Biggest Oil Reserves," *Bloomberg*, June 14, 2012, http://www.bloomberg.com/news/2012-06-13/venezuela-overtakes-saudis-for-largest-oil-reserves-bp-says-1-.html.

3 Joe Romm, "WikiLeaks peak oil bombshell: Saudi Arabian reserves overstated by 40%, global production plateau imminent," *Thinkprogress.org*, February 9, 2011, http://thinkprogress.org/climate/2011/02/09/207484/wikileaks-peak-oil-saudi-arabia-reserves-overstated/; Acil Tabbara, "Oil giant Saudi Arabia looks to alternative energy, *Phys.org*, January 25, 2011, http://phys.org/news/2011-01-oil-giant-saudi-arabia-alternative.html; also see Laura El-Katiri, "Energy Sustainability in the Gulf States: The Why and the How," *The Oxford Institute for Energy Studies*, March 2013, p. 9.

4 "Saudi Arabia, *EIA*, September 10, 2014. According to the CIA's Factbook on Saudi Arabia, oil accounts for 80 percent of budget revenue and 90 percent of export earnings. Central Intelligence Agency (USA), "Saudi Arabia," *The World Factbook*, https://www.cia.gov/library/publications/the-world-factbook/geos/sa.html.

5 *BP Statistical Review of World Energy*, June 2014.

6 In 2010 it spent over $30 billion on oil and about $12 billion on electricity subsidies. Its fossil fuel subsidies thus are the second largest in the world after Iran and larger than those of Russia, India and China, which follow it in the ranking. Eckart Woertz, "The Domestic Challenges in the Saudi Energy Market and Their Regional and Geopolitical Implications," *Norwegian Peacebuilding Resource Centre Policy Brief*, November 2013, http://www.peacebuilding.no/var/ezflow_site/storage/original/application/05a485f202440778052158eb7ef9808b.pdf; "Saudi Arabia," *Energy Efficiency Report*, p. 3.

7 Saudi Arabia," *EIA*, September 10, 2014.

8 *BP Statistical Review of World Energy*, June 2014.

9 *BP Statistical Review of World Energy*, June 2014.

10 Woertz, "The Domestic Challenges," November 2013.

11 These countries have 185.6 million people, seven times more than Saudi Arabia. Egypt's population is 79.39 million, Algeria's 37.76 million, Sudan's 36.43 million and Morocco's 32.06 million. Saudi Arabia has a population of 28.4 million. "Saudis use nine times more electricity than fellow Arabs," *Arab News*, February 19, 2014, http://www.arabnews.com/news/527936.

12 See *Saudi Arabia Investment and Business Guide* (Washington DC: International Business Publications, 2011) p. 79; "Saudi and GCC Opportunities, 2012-16," *Gulf in the Media*, June 2012, http://www.gulfinthemedia.com/files/article_en/608373.pdf, p. 15; "Summer demand taxes Saudi power sector, but kingdom is working on solutions," *International Energy Agency*, August 6, 2014, http://www.iea.org/newsroomandevents/news/2014/august/summer-demand-taxes-saudi-power-sector-but-kingdom-is-working-on-solutions.html.

13 The nation consumed 310,000 barrels a day of fuel oil in 2013, the most since 2008. Imports averaged 101,417 barrels a day up 67 percent from 2012 and the highest since at least 2002. Wael Mahdi, "Saudi Arabia Seen Needing More Fuel Oil to Produce Power," *Bloomberg*, February 27, 2014, http://www.bloomberg.com/news/2014-02-27/saudi-arabia-seen-needing-more-fuel-oil-to-produce-power.html; see also Cameron McHugh, *Arab News*, March 7, 2014, http://www.arabnews.com/news/536381.

14 For a discussion on this argument see, Mohammad Asif, Wasim Ahmad, Raj Bahadur Sharma, "Economic diversification in Saudi Arabia: A disaggregated analysis," *College of Business Administration* (CBAK), Salman bin Abdulaziz University, Al-Kharj-11942, Kingdom of Saudi Arabia (KSA), http://cbagccu.org/files/pdf/2/1.pdf. For a discussion on economic diversification in Saudi Arabia see, Martin Hvidt, "Economic Diversification in GCC Countries: Past Record and Future Trends," Research Paper, *Kuwait Programme on Development, Governance and Globalisation in the Gulf States*, Number 27, January 2013, http://www.lse.ac.uk/IDEAS/programmes/kuwait/documents/economic-diversification-in-the-gcc-countries.pdf.

15 Eckart Woetz, March 20, 2014; Bassam Fattouh, "Summer Again: The Swing in Oil Demand in Saudi Arabia," *The Oxford Institute of Energy Studies*, July 2013; Daniel Fineren, "Saudi Arabia may be overcoming addiction to oil-fired power," *Reuters*, November 18, 2013, http://www.reuters.com/article/2013/11/18/us-saudi-oil-demand-idUSBRE9AH0Q820131118.

16 See Caryle Murphy, "Saudi Arabia's Youth and the Kingdom's Future," Woodrow Wilson Centre for Scholars, Middle East Program, *Occasional Paper Series*, Number 2 (Winter 2011), http://www.wilsoncenter.org/sites/default/files/Saudi%20Arabia's%20Youth%20and%20the%20Kingdom's%20Future%20FINAL.pdf, p. 3.

17 "Construction Sector In The Kingdom Of Saudi Arabia," *Mordor Intelligence*, September 2014, http://mordorintelligence.com/product/construction-sector-in-the-kingdom-of-saudi-arabia-analysis-of-commercial-industrial-infrastructure-institutional-and-residential-construction-along-with-trends-over-the-last-decade/.

18 "Kingdom addresses domestic energy consumption head-on," *Saudi Gazette*, September, 16, 2012, http://www.saudigazette.com.sa/index.cfm?method=home.regcon&contentid=20120916136293.

19 Saudi: Oil importer by 2030? September 06, 2012, *Zawya* (UAE), http://www.zawya.com/story/Saudi_Oil_importer_by_2030-ZAWYA20120906051210/; Prachi Patel, "Solar-Powered Desalination," *MIT Technology Review*, April 8, 2010, http://www.technologyreview.com/news/418369/solar-powered-desalination/.

20 Cameron McHugh, "Demand for refined oil products to stay high," *Arab News*, March 7, 2014, http://www.arabnews.com/news/536381.

21 Glada Lahn and Paul Stevens, *Burning Oil to Keep Cool: The Hidden Energy Crisis in Saudi Arabia* (London: Chatham House, December 2011), 38p; Brianna Panzica, "When OPEC Fails...," September 9, 2012, *Energy and Capital*, http://www.energyandcapital.com/articles/when-opec-fails/2546.

22 Ayesha Daya and Dana El Baltaji, "Saudi Arabia May Become Oil Importer by 2030, Citigroup Says," *Bloomberg*, Sep 4, 2012, http://www.bloomberg.com/news/2012-09-04/saudi-arabia-may-become-oil-importer-by-2030-citigroup-says-1-.html.

23 Christopher Helman, "Saudi Arabia To Become An Oil Importer? Here's How They Can Avoid It," *Forbes*, September 4, 2012, http://www.forbes.com/sites/christopherhelman/2012/09/04/saudi-arabia-to-become-an-oil-importer-heres-how-they-can-avoid-it/.

24 Heidy Rehman, "Saudi Petrochemicals: The End of the Magic Porridge Pot?" *Citigroup*, London, September 2012, p. 1. For analysis of the report see, Ambrose Evans-Pritchard, "Saudi oil well dries up," *The*

Telegraph, September, 2012, http://blogs.telegraph.co.uk/finance/ambroseevans-pritchard/100019812/saudi-oil-well-dries-up/; "Saudi Arabia May Run Out of Oil to Export by 2030," *Energy Trends Report*, September 6, 2012, http://www.energytrendsinsider.com/2012/09/06/saudi-arabia-may-run-out-of-oil-to-export-by-2030/.

25 Sara Hamdan, "Demand Booms Among Saudis," *New York Times*, June 17, 2014 http://www.nytimes.com/2014/06/18/business/energy-environment/18iht-ren-saudi18.html?_r=0.

26 Heidy Rehman, "Saudi Petrochemicals," September 2012, p. 4; Also see Abid Ali, "Saudis may not have oil to export by 2030," *Aljazeera*, September 6, 2012, http://blogs.aljazeera.com/blog/middle-east/saudis-may-not-have-oil-export-2030.

27 Bill Spindle, "Break-Even Oil Price Bogeyman Stalks Gulf Economies," *The Wall Street Journal*, May 28, 2013, http://blogs.wsj.com/middleeast/2013/05/28/break-even-oil-price-bogeyman-stalks-gulf-economies/.

28 Angus McDowall, "Saudis say fuel subsidies distort economy, must be cut," *Reuters*, May 7, 2013, http://www.reuters.com/article/2013/05/07/saudi-energy-subsidies-idUSL6N0DO12P20130507.

29 Duncan Clark, "Fossil fuel subsidies: a tour of the data," *The Guardian*, January 19, 2012, http://www.theguardian.com/environment/datablog/2012/jan/18/fossil-fuel-subsidy.

30 Mauela Paraipan, "Saudi Arabia's role in the Broader Middle East," *World Security Network*, January 8, 2010, http://www.worldsecuritynetwork.com/Other/manuela-paraipan/Saudi-Arabias-role-in-the-Broader-Middle-East.

31 Lahn and Stevens, *Burning Oil*, December 2011, p. 15.

32 "Nuclear Power in Saudi Arabia," *World Nuclear Association*, May 2014, http://world-nuclear.org/info/Country-Profiles/Countries-O-S/Saudi-Arabia/.

33 Woertz, "The Domestic Challenges," November 2013.

34 See Rehman, "Saudi Petrochemicals," September 2014, p. 36.

35 William J. Broad and David E. Sanger, "With Eye on Iran, Rivals Also Want Nuclear Power, *The New York Times*, http://www.nytimes.com/2007/04/15/world/middleeast/15sunnis.html?pagewanted=all&_r=0, April 15, 2007.

36 Matthew Fuhrmann, *Atomic Assistance: How "Atoms for Peace" Programs Cause Nuclear Insecurity* (Ithaca, NY: Cornell University Press, 2012), pp. 136.

37 Mark Hibbs, "Saudi Arabia's Nuclear Ambitions," *Carnegie Endowment for International Peace*, July 10, 2010, http://carnegieendowment.org/2010/07/20/saudi-arabia-s-nuclear-ambitions/2fq6.

38 See, "The Nuclear Question in the Middle East: see nuclear energy in the Middle East," Centre for International and Regional Studies, Georgetown University School of Foreign Service in Qatar, *Summary Report*, Number 4; "Emerging Nuclear Energy Countries," *World Nuclear Association*, May 23, 2014, http://www.world-nuclear.org/info/Country-Profiles/Others/Emerging-Nuclear-Energy-Countries/.

39 Hibbs, "Saudi Arabia's Nuclear Ambitions," July 10, 2010.

40 Quoted in John P. Banks, Charles K. Ebinger (eds.), *Business and Nonproliferation: Industry's Role in Safeguarding a Nuclear Renaissance* (Washington DC: Brookings Institution Press, 2011), p. 86.

41 April Yee, "Saudi Arabia to seek bids for its first nuclear reactor", *The National*, November 11, 2013, http://www.thenational.ae/business/industry-insights/energy/saudi-arabia-to-seek-bids-for-its-first-nuclear-reactor.

42 "Trio team up for Saudi nuclear projects," *World Nuclear News*, July 12, 2010, http://www.world-nuclear-news.org/C-Trio_team_up_for_Saudi_nuclear_projects-1207104.html.

43 "Future of Nuclear Energy in Emerging Markets: Kingdom of Saudi Arabia", *Forum on Energy*, January 14, 2014, http://forumonenergy.com/2014/01/14/future-of-nuclear-energy-in-emerging-markets-kingdom-of-saudi-arabia/.

44 Daniel Fineren, "Nuclear Saudi Arabia a lifeline for the atomic energy industry," *Reuters*, April 23, 2013, http://uk.reuters.com/article/2013/04/23/uk-saudi-nuclear-idUKLNE93M00M20130423.

45 See Gawdat Bahgat, *Alternative Energy in the Middle East* (New York: Palgrave Macmillan, 2013), pp. 83-5.

46 "Teaming up for Saudi Bids," *World Nuclear News*, September 9, 2013, http://www.world-nuclear-news.org/NN-Teaming_up_for_Saudi_bids-0909137.html; "GE Hitachi, Exelon to explore nuclear energy opportunities in Saudi Arabia," *Electric Light and Power*, September 9, 2013, http://www.elp.com/articles/2013/09/ge-hitachi-exelon-to-explore-nuclear-energy-opportunities-in-saudi-arabia.html.

47 Exelon Nuclear Partners is a global provider of nuclear power operating, performance improvement, and new nuclear development services. It is the largest operator of nuclear power plants in the United States.

48 "A Delegation of Saudi Industry Representatives Visits France," *Areva*, November 26, 2013, www.areva.com; "Areva And EDF Sign two Series of Agreements with Companies and Universities for the Saudi Nuclear Program," *Areva Press Release*, December 30, 2013, www.areva.com; Kelvin Ross, "EDF and Areva sign Saudi nuclear pacts," *Power Engineering International*, January 06, 2014, http://www.powerengineeringint.com/articles/2014/01/edf-and-areva-sign-saudi-nuclear-pacts.html.

49 Summer Said, "Saudi Arabia, China Sign Nuclear Cooperation Pact, *Wall Street Journal*, January 16, 2012, http://online.wsj.com/news/articles/SB10001424052970204468004577164742025285500.

50 Eman El-Shenawi, "Saudi Arabia signs nuclear-energy deal with Argentina," *Al-Arabiya News*, June 29, 2011, http://www.alarabiya.net/articles/2011/06/29/155288.html.

51 "Saudi Arabia and Korea agree to cooperate," *World Nuclear News*, November 15, 2011, http://www.world-nuclear.net/info/Country-Profiles/Countries-O-S/Saudi-Arabia/; "Saudi Arabia, South Korea in nuclear cooperation deal," *Reuters*, November 15, 2011, http://www.reuters.com/article/2011/11/15/us-saudi-southkorea-idUSTRE7AE0GU20111115.

52 Quoted in Thomas W. Lippman, *Saudi Arabia on the Edge: The Uncertain Future of an American Ally* (Washington DC: Potomac Books, 2012), p. 60.

53 According to a Congressional Research Service (CRS) report, then-State Department Spokesperson P.J. Crowley coined the 'Gold Standard' term to describe the terms of the 123 Agreement between US and the UAE, wherein the latter foreswore the right to enrich uranium and process plutonium – processes crucial to building nuclear weapons. See Paul K. Kerr, Mark Holt, Mary Beth Nikitin, "Nuclear Energy Cooperation with Foreign Countries: A Primer," *Congressional Research Service Report for Congress (CRS Report)*, July 11, 2011, p. 17, http://fas.org/sgp/crs/nuke/R41910.pdf.

54 Under the NPT, signatories have a sovereign right to develop peaceful nuclear energy as defined by the UN International Atomic Energy Agency (IAEA).

55 For imperatives of America's cooperation with UAE and other Arab states see, Matthew Fuhrmann, *Atomic Assistance: How "Atoms for Peace" Programs Cause Nuclear Insecurity* (Ithaca, NY: Cornell University Press, 2012), pp. 135-37.

56 During the past several years, the United States' negotiation on nuclear cooperation has been under intense Congressional scrutiny. The US has attempted to persuade certain countries to forgo enrichment and reprocessing and conclude additional protocols. When UAE gave up ENR in its 123 agreement with the United States, President Obama argued that the agreement "has the potential to serve as a model for other countries in the region that wish to pursue responsible nuclear energy development." See *CRS Report*, July 11, 2011, p. 17.

57 "Saudi Arabia's nuclear programme," *Nuclear Engineering International*, August 8, 2013 http://www.neimagazine.com/features/featuresaudi-arabias-nuclear-programme/.

58 Daniel Fineren, "Nuclear Saudi Arabia a lifeline for the atomic energy industry," *Reuters*, April 23, 2013, http://uk.reuters.com/article/2013/04/23/uk-saudi-nuclear-idUKLNE93M00M20130423.

59 Yee, "Saudi Arabia to seek bids for its first nuclear reactor," *The National*, November 11, 2013.

60 Ian Jackson, "Nuclear energy and proliferation risks: myths and realities of the Persian Gulf," *International Affairs*, Vol. 85, No. 6, (2009), p. 1159.

61 See "Nuclear Power and Nuclear Weapons," *Nuclear Energy Information Service* (Illinois' Nuclear Power Watchdog), August 31, 2004, http://www.neis.org/literature/Brochures/weapcon.htm; Karl

Grossman, Nuclear Power/Nuclear Weapons -- and a Precarious Future," *Huff Post*, March 11, 2013, http://www.huffingtonpost.com/karl-grossman/nuclear-powernuclear-weap_b_2851985.html; "Outline History of Nuclear Energy," *World Nuclear Association*, March 2014, http://www.world-nuclear.org/info/Current-and-Future-Generation/Outline-History-of-NUclear-Energy/.

62 Richard Weitz, "Gulf Cooperation Council Moves Forward with Nuclear Energy Plans," *WMD Insights*, April 2007, http://www.nti.org/country-profiles/saudi-arabia/nuclear/.

63 Hibbs, "Saudi Arabia's Nuclear Ambitions," July 10, 2010.

64 Ewen MacAskill and Iyan Traynor, "Saudis consider nuclear bomb," *The Guardian*, September 18, 2003, http://www.theguardian.com/world/2003/sep/18/nuclear.saudiarabia; Simon Henderson, "Toward a Saudi Nuclear Option: The Saudi-Pakistani Summit," *The Washington Institute*, October 16, 2003, http://www.washingtoninstitute.org/policy-analysis/view/toward-a-saudi-nuclear-option-the-saudi-pakistani-summit.

65 Jackson, "Nuclear energy and proliferation risks," 2009, p. 1159.

66 Lippman, *Saudi Arabia on the Edge*, 2012, p. 245; Kate Amlin, "Will Saudi Arabia Acquire Nuclear Weapons?" August 1, 2008, *Nuclear Threat Initiative*, http://www.nti.org/analysis/articles/will-saudi-acquire-nuclear-weapons/; Gawdat Bahgat, "Nuclear Proliferation: The case of Saudi Arabia", *Middle East Journal*, Vol. 60, No. 3 (Summer 2006) p. 425; Richard Russell, "A Saudi Nuclear Option," *Survival*, Vol. 43, No. 2 (Summer 2001), p. 72.

67 Russell, "A Saudi Nuclear Option," Summer 2001, p. 70.

68 "Saudi Arabia: Country Profile," *Nuclear Threat Initiative (NTI)*, November 2013, http://www.nti.org/country-profiles/saudi-arabia/delivery-systems/.

69 "Saudi Arabia Bought Chinese Missiles With U.S. Backing: Report," *Global Security Newswire (NTI)*, January 30, 2014, http://www.nti.org/gsn/article/us-backed-secret-saudi-missile-purchase-china/.

70 Lipmann, *Saudi Arabia on the Edge*, 2012, p. 246.

71 Zachary Keck, "No, Pakistan Won't Sell Saudi Arabia Nukes," *The Diplomat*, August 14, 2013, http://thediplomat.com/2013/08/no-pakistan-wont-sell-saudi-arabia-nukes/.

72 Jackson, "Nuclear energy and proliferation risks," 2009, p. 1157.

73 Yana Feldman, "Saudi Arabia Country Profile: Nuclear Facilities Profiles," *Stockholm International Peace Research Institute*, July 2004, http://archives.sipri.org/contents/expcon/cnsc1sau.html.

74 Jackson, "Nuclear energy and proliferation risks," 2009, p. 1157.

75 Thomas W. Lippman, "Nuclear Weapons and Saudi Strategy," *Middle East Institute*, Policy Brief No. 5, January 2008.

76 Kate Amlin, "Will Saudi Arabia Acquire Nuclear Weapons?" August 1, 2008.

77 Richard Nield quoted in Naser al-Tamimi, "Will Riyadh Get the Bomb? Saudi Arabia's Atomic Ambitions," *Middle East Quarterly*, Vol. XX, No. 2 (Spring 2013), http://www.meforum.org/3509/saudi-arabia-nuclear-bomb#_ftnref57.

78 Christopher M. Blanchard, "Saudi Arabia: Background and U.S. Relations," February 12, 2014, *Congressional Research Service*, http://www.fas.org/sgp/crs/mideast/RL33533.pdf, p. 7.

79 Robert Mason, "Establishing equilibrium in the Gulf: in search of a pragmatic agenda for stability and security," *openDemocracy*, May 7, 2014, www.opendemocracy.net.

80 "Israel, Saudi Arabia Unite against Iran," *Al-Manar*, November 17, 2013, http://www.almanar.com.lb/english/adetails.php?fromval=1&cid=31&frid=31&eid=121251.

81 Jason Burke, "Riyadh will build nuclear weapons if Iran gets them, Saudi prince warns," June 29, 2011, *The Guardian*, http://www.theguardian.com/world/2011/jun/29/saudi-build-nuclear-weapons-iran.

82 Bradley Bowman, "Chain Reaction: Avoiding a Nuclear Arms Race in the Middle East," *Report to the Committee on Foreign Relations*, US Senate, February 2008, p. 11.

83 Abdulaziz Sager quoted in Tamimi, "Will Riyadh Get the Bomb? Spring 2013.

84 Lewis A. Dunn, "Strategic Reassurance if Iran "Goes Nuclear": A Framework and Some Propositions," *Strategic Insights,* Volume VIII, Issue 5 (December 2009), http://calhoun.nps.edu/public/bitstream/handle/10945/11160/dunnDec09.pdf?sequence=1.

85 Mai Yamani, "The Bewildered Kingdom," *Project Syndicate*, January 03, 2013, http://www.project-syndicate.org/commentary/mai-yamani-assesses-saudi-arabia-s-increasingly-assertive-regional-foreign-policy-since-the-start-of-the-arab-spring.

86 "Saudi Arabia May Go Nuclear Because of Obama's Iran Deal," *The Daily Beast*, February 14, 2014, http://www.thedailybeast.com/articles/2014/02/14/saudi-arabia-may-go-nuclear-because-of-obama-s-iran-deal.html.

87 Quoted in "European Oil Embargo Plans Seen Hitting Iran," *Global Security Newswire* (NTI), February 10, 2012, http://www.nti.org/gsn/article/european-oil-embargo-plans-seen-hitting-iran/.

88 Lipmann, *Saudi Arabia on the Edge*, 2012, p. 244.

89 Bruce Riedel, "Saudi Arabia puts on show with 'Abdullah's Shield'", April 30, 2014, *Al Monitor*, http://www.al-monitor.com/pulse/originals/2014/04/saudi-arabia-military-show-signal-washington-tehran.html.

90 Feroz Hassan Khan, *Eating the Grass: The Making of the Pakistani Bomb* (New Delhi: Foundation Books, 2013), p. 172.

91 Julian Borger, "Does Pakistan have nuclear weapons ready to ship to Saudi Arabia?" November 7, 2013, *The Guardian*, http://www.theguardian.com/world/julian-borger-global-security-blog/2013/nov/07/pakistan-saudi-nuclear-proliferation. At the beginning of 2014, Foreign Minister Saud Al-Faisal and Crown Prince Salman visited Islamabad to renew the two countries' military agreements on joint arms production. The visit was also intended to lay the groundwork to bring 30,000 Pakistani soldiers and military advisers to the kingdom. See Mai Yamani, Saudi Arabia's diplomatic pilgrimage to Pakistan, *The Japan Times*, March 28, 2014, www.japantimes.co.jp.

92 Tamimi, "Will Riyadh Get the Bomb? Spring 2013.

93 Mark Urban, "Saudi nuclear weapons 'on order' from Pakistan," *BBC News Middle East*, November 6, 2013, http://www.bbc.com/news/world-middle-east-24823846.

94 Thomas W. Lippman, "Saudi Arabia's Nuclear Policy," *Saudi-US Relations Information Service (SUSRIs)*, August 5, 2011, http://susris.com/2011/08/05/saudi-arabia's-nuclear-policy-lippman/.

95 Zachary Keck, "Pakistan's Energy Savior: Iran," *The Diplomat*, February 27, 2013, http://thediplomat.com/2013/02/pakistans-energy-savior-iran/.

96 Lippmann, *Saudi Arabia on the Edge*, 2012, p. 246.

97 Arjun Makhijani, *Carbon-free and Nuclear-free: A Roadmap for U.S. Energy Policy* (Muskegon, MI & Tahoma Park, MD: RAR Books & IEER Press, 2008), p. xix.

98 "Kingdom renews its call to make Middle East a zone free of nuclear weapons," *Al-Riyadh*, April 30, 2014, http://www.alriyadh.com/en/article/931706/Kingdom-renews-its-call-to-make-Middle-East-a-zone-free-of-nuclear-weapons.

99 John Glaser, "Saudi Official Calls for Nuclear Weapons-Free Zone in Mideast," *Antiwar.com*, January 25, 2012, http://news.antiwar.com/2012/01/25/saudi-official-calls-for-nuclear-weapons-free-zone-in-mideast/.

100 Patrick W. Ryan, "Middle East Nuclear Proliferation," *SUSRIS*, November 8, 2013, http://susris.com/2013/11/08/middle-east-nuclear-proliferation/.

101 "Middle East WMD-Free Zone," *British American Security Information Council*, December 2013, http://www.basicint.org/sites/default/files/basic_mewmdfz_briefing_dec2013_0.pdf.

102 "The Middle East WMD Free Zone," *International Law and Policy Institute*, February 13, 2014, http://nwp.ilpi.org/?p=1225.

103 Michael Pizzi, "Obama doctrine under scrutiny in an anxious Saudi Arabia," *Aljazeera America*, March 27, 2014, http://america.aljazeera.com/articles/2014/3/27/obama-saudi-visit.html.

104 Robert E. Hunter, *Building Security in the Persian Gulf* (Santa Monica, California: RAND, 2010).

7

Legal Matters Impacting the Oil and Gas Industry in the Gulf Cooperation Council (GCC) Countries: Keeping Legal Reform and Regulatory Frameworks Current in a Changing Energy Landscape

Mary B. Ayad[1]

Background and Industry Relevance

Why are oil[2] and gas concessions and agreements still relevant in the changing energy landscape? The author submits that Middle East oil is not dead. Here is why. A recent article: 'The US Shale Oil Boom: Potential Impacts and Vulnerabilities of an Unconventional Energy Source', a policy briefing from Harvard Kennedy School written in June of 2013 states that the US may become the biggest oil producer in the world in a few years. This will drive OPEC oil prices down and create economic problems in OPEC countries who consume as much oil as they export. Their oil income is losing value, causing a greater gap in funds for their low cost subsidies. This will lead to a need to develop infrastructure for renewable energy sources. This means high risk contracts. Adjudicatory risk in energy contracts is going to be a key issue. The legal principles and lessons learned from oil concession disputes are applicable to renewable energy agreements. The history of oil concessions led to mistrust on the side of Arab parties but Gulf Cooperation Council states like Saudi Arabia which has recently reformed its international arbitration law to make it legally binding offer significant hope. The United Arab Emirates will also be highlighted. These legal reforms will support an improved regulatory framework that will facilitate renewable energy agreements. An understanding of the different types of oil agreements such as production sharing, joint ventures or concessions means a better understanding of the differing types of risks. Some types of contracts have higher risks than others and this will be exacerbated with inherently higher risk renewable agreements.

Changes in global financial markets and the energy industry offer major challenges and opportunities for the Gulf Cooperation Council states for decades to come. Changes in energy are intertwined with changes in global financial markets. The shift in relying on shale oil from the United States is the world's most important matter in the energy sector, cutting across multiple legal disciplines, requiring expert advice. Shale oil exports by the US will increase trade agreements and concessions between the US and importing countries. Matters related to infrastructure projects will require excellent pre-negotiation and contract drafting. Shale oil exports from the US will drive down OPEC prices creating a need for Gulf Cooperation Council states to develop massive infrastructure projects to sustain renewable energy projects. Major world economies will certainly be impacted, thus affecting the global financial market. Adjudicatory risk in renewable projects and more technologically advanced oil projects will remain an important topic for commercial lawyers as will international arbitration and accurate analysis of regulatory frameworks for investor clients, whether they are private investors or government parties to a contract. The changing climate in the Middle East requires lawyers who understand those legal climates. Lawyers must be updated on the rapid reforms to investment laws, free zones and arbitration courts, and international arbitration laws and understand that even within the same country there are different jurisdictions as in the case of Dubai, which has different laws from the United Arab Emirates (UAE). Firms with lawyers already well-versed in the MENA will have an edge and this means a growing market for lawyers from and based in the Gulf Cooperation Council states. Counsel and parties to international contracts, especially in big oil or renewable energy agreements need to understand specific doctrines related to certain legal questions. These questions include (1) what arbitration rules one should choose (ICC[3], LCA,[4] HKIAC,[5] DIAC,[6] etc.), (2) what seat one should select,[7] (3) how does the arbitration clause interact with other provisions in the contract, and (4) should provisions be made for consolidation or joinder of claims or sole party. The latter is especially relevant in production sharing. The importance of the seat is that it determines the law, or *lex arbitri*. Therefore care is needed. For example the arbitration outcome in Dubai is different than that of the UAE[8] because although Dubai is located in the UAE –they have different jurisdictions and different laws, as will be discussed in the section on arbitration. The matter of pursuing a claim where assets are held should not cloud judgement in the choice of seat. Public policy, sovereign immunity, and competence challenges emerge as defences in the nexus of the *lex arbitri* and international arbitration.

This research focuses on the changing global energy landscape[9] as it impacts the Gulf Cooperation Council states and its strategic implications with respect to large-scale commercial contracts (involving various types of energy: mineral, petrol, coal, natural gas, shale and fracking). The reasons the oil and gas industries are still highly relevant will be extrapolated in the following discussions. A major component falling under the category of 'strategic' includes a strategic policy to hedge unnecessary risk and to create a minimal risk environment for investors.[10]

The risky nature[11] of renewable[12] energy projects and contracts involves substantial sums and substantial potential losses. For example, here is a case in point:

> 'In 2008 alone, Spain was responsible for half of the world's solar power facilities in terms of wattage. Unfortunately, the financial downturn in late 2008 had major repercussions for Spain, as its solar industry plummeted. Madrid would soon face a growing deficit in its energy budget of around €4.5 billion (US$6.1 billion) by 2012. As a result, Spain, once the crown jewel of the solar power industry, gutted its market by removing its FITs, cutting subsidies and capping the rate of return for investors.'

The disastrous impact of cutting subsidies had repercussions not just on the economy but on relations with foreign partners.

> 'Needless to say, these measures have incensed the thousands of PV investors who will now see little benefit in their ventures in Spain. What has followed has been a seemingly endless carousel of lawsuits directed towards the Spanish government, as PV investment, development and installation companies such as the Foresight Group and European Energy have lined up to launch grievances against Madrid's reforms and subsequent cuts. Yet another major PV player entered into the fray in February 2014, as a subsidiary of UAE-backed Company Masdar filed a legal complaint against Spain before the International Centre for Settlement of Investment Disputes (ICSID) in order to protect its investments from these reforms. Currently, the ICSID has five cases against Spain on record, four of which revolve around renewable energy.'

The consequences of the risks of these projects cannot be taken lightly. Indeed:

> 'Global energy and electricity demands are set to grow for decades. A growing world population and economic development predict faster grow rates for electricity than for primary energy demand, with most growth in developing countries. The UN population projections estimate population growth from 6.8 million people in 2010 to 8.365 billion at 2025 and 10.633 million in 2050.'[13]

Nuclear projects are a potential renewable source of energy that can fill the gap that non-renewable sources are predicted to create. The MENA region has a substantial contribution toward this end with $400 billion worth of nuclear projects planned.[14]

The author is of the view that oil and gas contracts and agreements are still highly relevant for the very reason that lessons learned from these types of contracts are transferable and applicable to future renewable energy contracts and especially to infrastructure and construction projects as these will need to be further developed to support sustainable and renewable energy contracts.[15] Not only are prior contract disputes important learning tools but improving the extant regulatory framework in oil and gas and other mineral contracts will facilitate the regulatory framework for renewable sources, especially nuclear energy.[16]

In order to demonstrate the need for hedging risk a thorough analysis of the different types of contracts and the different types of risk they incur is given.

General Matters in the Oil and Gas Industry

There are economic implications to the US policy on shale. From an economic perspective the increased supply of shale means a lower demand for oil and the pushes the price down. However: 'Conventional wisdom says if the United States drops its oil imports to 25 percent of demand, the oil will come from North American sources. This scenario is price dependent. If the marginal price of oil drops, the cheapest oil will be from the Middle East, and oil from Venezuela and Canada will be more expensive.'[17]

The possibility of the US dropping its oil imports is feasible: 'An analysis of more than 4,000 shale wells along with the activities of one hundred oil companies involved in shale oil exploitation suggests that the united states may become the largest global oil producer in just a few years.'[18] A number of opportunities abound with respect to the oil and gas industry. These opportunities are a necessity due to the changes in the energy[19] industry and the economic ramifications. Before discussing the opportunities the author suggests the reforms to the regulatory framework can be enhanced through the doctrine of public interest, or in Arabic '*maslaha*'. This is a valid doctrine of Islamic jurisprudence and can support legal reform. Although reform will be discussed in later sections of this paper, the author suggests that pre-existing models that incorporate the concept of public interest can be useful for drafters of similar legislation in MENA countries. The US can serve as a good model globally for encouraging renewable energy. On the basis of public interest (equal to the doctrine of *maslaha*). An example from the Energy Law of the State of New York is as follows:

> It is declared that the cost of electricity is an essential factor in the state's economy, that energy efficiency improvements can reduce the cost of electricity, that the authority's energy efficiency projects can reduce the cost of electricity for its customers, and that it is in the public interest for the authority to utilize petroleum overcharge restitution funds to undertake energy efficiency and renewable energy projects for public and private entities.[20]

Thus, it should be the policy of GCC countries to put funds toward renewable energy and this can be supported by the doctrine of *maslaha* at Islamic law. The Gulf Cooperation Council states need a stronger energy policy and law specifically dealing with energy. The US energy law can offer drafters a working model.[21] It is a matter of public interest[22] for the government to maximize energy usage and minimise waste either by establishing energy optimising domestic infrastructures or through developing infrastructure and usage of renewable and non-renewable energy sources. The date of the US energy law is equally important. As far back as 1976 the US was considering renewable energy sources. It was after a major energy emergency crisis. The US can certainly do better in developing more solar and wind infrastructure and resources. The value of the US experience to GCC states is to plan in advance and not wait until an energy shortage crisis takes place, i.e., not to wait as long as the US waited to build up a better infrastructure for renewables, especially solar and wind power. The US passed laws to support the earlier policy mentioned in Article 1. This regulatory framework deals specifically with energy matters, the energy law and includes provisions as the Fuel Set-aside Act (Article 10), Conservation Construction Code (Article 11), Solar Energy Products Warranty Act (Article 12), Energy Supply and Production Act (Article 21), and Temporary Waste Repositories (Article 18) that were drafted at an early stage and essential before renewable projects were begun. Drafting similar laws helps create a solid and stable regulatory framework, one that is transparent, reliable and attractive to investors and one that reduces the perception of risk.

The time to act is now: while there is sufficient capital and remaining non-renewable energy supplies that *still* have market value.

Amongst the many opportunities, currently three key geopolitical factors are acting as strong opportunities for the GCC Countries. They are (1) America's new policy to lesson dependence on foreign oil, (2) Europe's capital crisis and (3) the Arab Spring. The United States is currently developing and implementing new drilling techniques which have boosted domestic production of both oil and natural gas, to trim imports. This means they will need experts from the MENA and the GCC (such as Saudi and Gulf Arab engineers) who have experience in oil drilling to share technical expertise. This could create opportunities for cooperation with a reversal of the industry, in that the technical expertise can come from engineers and scientists with exploration taking place in the United States. This means greater opportunity for business deals and contract between the US and oil producing Arab states.

Outside of the legal implications from an economic and investment point of view, the combination of the three factors noted above is that other nations, in nearer physical proximity to Gulf Cooperation Council States can be recipients of imports and investment. This creates opportunities. The impact of the Global Financial Crisis on the EU has lessened its capital.

The implication of this for the Gulf Cooperation Council States is this. One of these opportunities lies in the fact that the Eurozone is primarily a demand centre with limited production within its borders. This can strengthen regional economic development. The author submits that it could not be a better time for increased regional cooperation. This will be discussed further in the concluding paragraph with respect to regional treaties. Another source of these opportunities lies in the fact that the Eurozone is primarily a demand centre with limited production within its borders.[23] This means that Europe must look outside its borders to obtain its oil and gas needs.[24] Long-term energy demands in Europe are stable.[25] Longer-term oil and gas demand is driven by non OECD country growth and emerging markets.[26] All of this is largely unaffected by the European financial crisis. However, there is a reduction in availability of capital in that oil and gas companies face the tightening of capital and credit markets inside the Eurozone, as well as externally.[27] The implication of this is that companies may have to look beyond their traditional European banking partners to meet their capital needs.[28] The legal implication of this is that there is a rising supplier and counterparty risk assessment in which a higher need for identification of third party risk exposure, and possible renegotiation of contract terms are foreseeable factors.[29] In addition to opportunities a number of challenges exist. These are largely to do with taxation, corruption and indemnification. A second factor is that of the change of US policy from exploring petrol overseas to exploration in the United States with a stronger focus on locating domestic gas and petrol sources.

The author submits that the Arab Spring initially led to positive reforms in updating investment and arbitration laws. Although the Arab Spring has frozen, positive reforms can still take place. Greater transparency, updated investment codes, updated international arbitration laws, more free zones and more arbitration courts can enhance the regulatory framework to support large-scale non-renewable and renewable projects.

Economic Implications on Price

There are economic implications to the US policy on shale:

> 'OPEC members are most troubled by the downward pressure on oil prices from booming North American supply and softening global demand. The Wall Street Journal has summarized the problem. Each OPEC member has a "break even" price at which oil must sell for that country to balance its budget. In the past decades, these break-even prices rose rapidly as OPEC countries directed oil revenues towards massive investments in infrastructure projects and, more recently in expanding social programs meant to stave off upheaval cause by the Arab revolutions around them. The higher spending of OPEC countries has left them vulnerable to a drop in prices.'[30]

From an economic perspective the increased supply of shale means a lower demand for oil and this drives the price down. However, a factor that adds to the relevance of Middle East oil is this:

> 'Conventional wisdom says if the United States drops its oil imports to 25 percent of demand, the oil will come from North American sources. This scenario is

> price dependent. If the marginal price of oil drops, the cheapest oil will be from the Middle East, and oil from Venezuela and Canada will be more expensive.'[31]

The potential for the lowered prices is a likely reality due to the fact that 'an analysis of more than 4,000 shale wells along with the activities of one hundred oil companies involved in shale oil exploitation suggests that the United States may become the largest global oil producer in just a few years.'[32] The main factors driving this are improvements in technology and cost.[33]

Technical Differences in Types of Contracts

Within the oil and gas industry a number of varieties of contracts exist. The standard arrangements are licenses, production sharing contracts, joint ventures and service contracts, confidentiality agreements, study and bidding agreements, operating agreements, farm-out agreements, lifting agreements and gas sales contracts, and joint development agreements. Only the first three will be covered in this paper. Irrespective of the kind of arrangement or structure of a project it is important to note the following, '[w]here the value of the project has been substantially reduced (or destroyed) by the measures, the investor may have a claim for indirect expropriation. Indeed, the most common approach is for the investor to claim for both indirect expropriation and violation of 'fair and equitable treatment'. In practice, this is the combination that is most often used to respond to host state measures, including those that fall under the banner of resource nationalism.'[34]

Administrative Contract

It has been argued with merit that oil concessions are a different type of contract from other contracts: 'A "concession agreement" is an instrument between a state and a private person and providing for the granting by the state to the individual of certain rights or powers which normally would belong to and be exercised by the state. Typical examples of concessions are those relating to public utilities and the exploitation of natural resources.'[35] The author is of the view that administrative contracts are by their nature different from non-administrative contracts by virtue of the nature of one of their contracting parties. Notwithstanding this, they are also commercial contracts. Their dual nature requires technical expertise in drafting negotiations and contracts dealing with natural resources. The regulatory framework of a system of domestic laws, international arbitration laws and extant international instruments has to be able to support the dual nature of these contracts to the satisfaction of both parties equitably. This fact applies to all contracts but more so to Production Sharing Agreements, Joint Ventures, and Oil Concessions and Licenses discussed in the next sections.

Production Sharing Agreements

The essence of the Production Sharing Agreements (PSAs) rests on the premise that it is an administrative contract and on the principle that the state owns the natural resources.[36] Additionally, PSAs are complex. They are legally [37] and fiscally complex because unknown variables in future profits are a concern and because cost recovery for current operating expenses and capital investment expenditures do not fall into easy black and white categories.[38]

> 'Under a PSA, an oil company carries most of the financial risk of exploration and development. The state also faces some risk. Often the national oil company joins the consortium as an interest holder in the PSA, contributing some of its profits as "share capital" to the consortium that is developing the area granted by the PSA. Often the host government has the cost of its initial contribution "carried" by the other companies. This carried cost will be repaid to the companies from the host government's future profits under the PSA.'[39]

In short, 'the state would retain ownership of the resources and negotiate a profit-sharing system.'[40]

The author is of the view that the higher burden of risk lies upon the commercial party contracting with the state and that these types of contracts are more beneficial to the state as long as they are willing to accept the higher risks involved with calculating profit considerations. The foregoing is true so long as the state does not expropriate. The balance of risk of loss is higher for the oil company.[41] Of all the various types of contracts, the author is of the view that when governments are facing consideration of risk when entering into renewable energy contracts, Production Sharing Agreements offer substantial protection in comparison to other types of agreements. This is if it remains a contract and is not enacted into law. Enacting it as law incurs legal consequences and is not conducive to the investment environment.[42] One matter to keep in mind with respect to production sharing agreements in the energy sector is

> [r]esource nationalism can take the form of measures that purport to modify the terms of the foreign investor's participation in a particular project. For example where a foreign oil company is in a production sharing arrangement with the host state, the latter may enact measures that adjust or tighten the cost recovery regime under Production Sharing Contract (PSC), thereby reducing the foreign participant's share of the off-take. Indeed, measures that interfere with the cost recovery rules of PSCs are an increasingly common form of resource nationalism in the oil industry.[43]

Another source of recourse for a foreign investor against sovereign risk is found in the ACIA (discussed elsewhere in this article) whereby:

> The ACIA definition of 'investment' therefore captures the instruments and interests typically associated with an oil and gas project. It is important to note that, under this definition, the instruments that underpin the investment are themselves investments. To apply this to the cost recovery example of resource nationalism provided above, where the cost recovery clause of a Production Sharing Contract (PSC) is unilaterally modified by the host state (or the contracting party that represents it), the foreign investor may be able to resist or seek compensation for that measure through a claim for indirect expropriation or breach of 'fair and equitable treatment'. This is because the PSC qualifies as an "investment'.[44]

Joint Ventures and Service Contracts

Joint ventures (JVs) do not fall under a generally understood definition.[45] At best, 'a JV simply implies that two or more parties wish to pursue a joint undertaking in some still to be clarified form.'[46] Joint Ventures are less common than oil agreements.[47] Joint Ventures contain multiple unknowns.[48] The author recommends when joint ventures either for oil concessions or renewable energy sources are entered into, competent counsel on both sides be involved in the pre-negotiation and drafting of these agreements because renewable energy contracts are already invoking greater elements of the unknown. The advantage of choosing a joint venture is that the host government is not alone in the decision making or responsibility of the project[49] however this is also the disadvantage because it means risks and costs are shared.[50] This makes the government a responsible party in the natural resource extraction and therefore potentially liable, including for environmental damages.[51] This is a higher risk contract and past lessons learned from oil and gas cases of joint ventures can be applicable to renewable energy particularly due to the risks involved in a speculative venture. However, due to the fact that they are infrequent, oil concessions and licenses serve as a better prototype for renewable energy contracts.

Licenses

An oil concession may also be referred to as a license agreement. In its essence, this type of agreement grants an oil company (the commercial actor or investor) rights to explore, develop, sell, and export oil and minerals extracted from a precise location for a fixed duration.[52] These rights are exclusive and competitive.[53] These types of agreements are common across the globe.[54] These types of agreements can be construed as administrative contracts and can activate substantial legal doctrines. These types of contracts are relatively straightforward, and less complex than joint ventures or production-sharing agreements.[55] The adjudicatory risk of these contracts comes from gaps in 'an acceptable and reliable legal infrastructure, including a judiciary capable of interpreting complex agreements'[56] since without these elements in the regulatory framework, 'the burden of devising contractual provisions to fill in gaps in the legal system of the host country'[57] is a serious consideration. The adjudicatory risk pertaining to concessions will be dealt with in the next section. The commercial risk has to do with the speculative nature of the contract because

> '[t]here is normally a lack of adequate knowledge about the potential of a concession area because seismic exploration has not been fully undertaken. The result is that the bidding system is often simply an action. Oil companies have no choice but to take calculated risks about what price to bid for a license. A company will be cautious in the amount it is prepared to bid since there is no guarantee the concession will cover the company's costs.'[58]

These types of concessions or licenses may be less replicable to renewable sources since they are confined to precise locations, however the legal aspects concerning how these types of disputes have been resolved as certainly relevant to other contracts in the renewables category. Additionally, the factor of calculated risk is relevant because with all energy projects there is no foreknowledge of the precise return of profit.

Future Applicability

Oil contracts and agreements are still vital and substantial. For example, just last year the United Arab Emirates (UAE) led the MENA region in high value contracts in the oil and gas sector as a party with the Zakum Development Co (ZADCO) awarding them a 3.7 billion contract to develop the offshore upper Zakum field.[59] In fact, in the second quarter of 2013 a total of 6.7 billion worth of engineering, procurement and construction (CEPC) deals were awarded of which the ZADCO contract represented 50% in value.[60] These types of contracts demonstrate the integration of innovation and technology to optimise oil production and sustainability with new and expanding projects.[61] The integration of oil and gas project infrastructure with technology is an important component of success,[62] but more than that, this trend is a replicable prototype for other types of energy infrastructure integrated with technology. This trend paves the way for future similar projects in renewable energy. There are similar contracts involving the UAE[63] and Qatar.[64] Another trend is the rehabilitation of ageing oil and gas fields throughout the region.[65] Again this is as a result of the marriage of technology and innovation with the energy industry. There is enormous potential for contractors looking to work on rehabilitation upgrade and modification schemes in the GCC.[66] This is a high potential market.[67] The author submits that this potential market has relevance to the oil and gas industry for the next foreseeable fifteen to thirty years at least due to the long terms duration of investor-state and investment contracts that may extend beyond this predicted time frame.[68] In addition, the industry and legal lessons learned from these types of contracts are applicable to renewable energy contracts particularly due to the doctrines and provisions dealing with public policy, administrative contracts and international public law.

Brief History of the Early Gulf Arab Oil Arbitrations

From a legal perspective, the history of oil and gas concessions in the Gulf is largely summed up by the early oil concessions in which:[69] 'In practice, especially the legal regime of oil and gas projects by multinational companies has been determined in large part by investment agreements. In the decades before 1945, these agreements (concessions) typically covered large areas of land, transferred title of the oil reserves to the investor, and did not contain an obligation of the investor to explore or produce oil. Under these agreements the host country would receive a bonus for the concession as such and royalties for barrels actually produced.'[70]

Early Oil Concessions

It is of the author's view, concurrent with other scholars,[71] that the royalties received did not reflect a fair price for the oil in light of rising oil costs on the market during the duration of the long term concessions. The contracts ought to have been renegotiated on the basis of equity. To an extent the legal history attempted to evolve with keeping to the ethos of the times: 'A second generation of agreements emerged in the 1960s and the 1970s. These reflected the new power of oil-producing countries, their desire to control their resources, and their aspirations to develop the necessary skills and technologies within their own borders. Often, state-owned companies were set up for the purpose of concluding and supervising the agreements with the foreign investor.'[72] With this positive development came more risks and yet also more opportunity in that: 'Areas with potential reserves were more restricted and closely defined, and the title to oil and gas remained with the host country. The risks inherent in a failure to find suitable oil and gas were shifted to the foreign investor who was, however, allowed to recover exploration costs in case commercial usable reserves were found.'[73]

> The early history of Gulf Arab Oil concessions was unfavourable to the Arab States and was followed by nationalisation of the oil industry after World War II, across the Middle East and North Africa (Algeria, Iran, Iraq, Libya, Kuwait and Saudi-Arabia.)[74]

Second Generation Contracts

The second generation of contracts opened the door to improved legal standing for both parties: 'Beyond the allocation of rights, tasks, risks, and responsibilities, investment contracts had to lay down the ground rules on which the parties agreed. These rules included, in particular, the law applicable to the project and the choice of a forum for dispute resolution. Specific provisions were often included concerning force majeure, good faith and changed circumstances. From a legal perspective, the most complex and difficult questions often concerned the inclusion of clauses regulating the conduct of the parties in case of political changes in the host country and in case of changes in the economic equilibrium between the host state and the investor.'[75]

Impact of Gulf Arab States

The current regulatory, legal and investment climate in the Gulf Arab States is extremely positive. The former Arab Spring reforms that took place throughout the MENA commercial legal environment beginning in 2011 (in spite of any growing pains) have led to increased investor confidence and increased investment. One example of this is the Saudi Arabian railway project. As far back as 2010 the Gulf Arab countries were planning a large scale railway project in which plans to build a national railway would start in one country and eventually link the Gulf Arab States to each other, with the possibility of continuing on to Europe. A project of this size and scale reflects tremendous confidence and optimism on a number of levels, from both the European and Gulf Cooperation Council sides as well as with respect to a growing appreciation for arbitration. In 2011 Renfe, the Spanish government owned railway company signed a $10 billion contract with the Kingdom of Saudi Arabia to build the first ever train in the Kingdom. The Spanish government was in competition with France but won the contract. It is understood that this contract will attract future similar contracts[76] but above all it represents the trend of increasing trust and investor confidence. Under the terms of this contract Spain qualifies as an 'investor' and the contract will bring about greater economic growth and prosperity to both sides.

Updates to International Commercial Arbitration and Investment Laws in the Gulf Arab States

It is helpful to remember that arbitration grew from the region as an indigenous and legitimate form of dispute resolution, beginning with the early arbitrations of the Prophet Mohammed at Banu Qurazyah and in the Quranic Verses dealing with arbitration of marriage disputes.[77] Thus, arbitration as an indigenous dispute resolution method plays an important intangible role in facilitating business relationships leading to greater economic growth and stability in the GCC States.

Four Gulf Arab states with remarkable regulatory frameworks and investment opportunities are highlighted in this paper. They are the Kingdom of Saudi Arabia, the United Arab Emirates (UAE), Iraq (including Kurdistan) and Kuwait. Other MENA and Gulf Arab countries are equally important but the focus here is also on recent reforms and investment-attractive updates that are extremely important from an investment perspective. It is also noteworthy to mention that Qatar has a new law, and although it remains a draft, there is hope that it will be enacted soon.

Investment Opportunities in the United Arab Emirates (UAE)

The nature of International Commercial Arbitration as a viable dispute resolution method is arguably the main reason for its success in positively impacting business relations and economic growth in the GCC States and especially in the United Arab Emirates. International Commercial Arbitration through its inherent nature provides for confidentiality allowing commercial parties to maintain their business reputations in the face of a dispute and allows them to maintain non-adversarial business relations as well as saving face, and this is a highly important cultural asset particularly in light of Middle Eastern business cultures and the reality of investor confidence when dealing with stakeholders who may otherwise react negatively if a commercial party in which they held interests was seen to be litigating or facing charges of breach of contract or financial losses. Thus, the confidence of investors and commercial parties in the Dubai institutional arbitration centers of investors and commercial parties is evidenced by the fact that many of the top- tier and globally oriented commercial law firms are highly interested in operating in the UAE as a result of the transparency of the newly reformed Dubai Laws. The value of free trade zones and updates to laws will have benefits in attracting investors for renewable energy contracts as well as the current construction, finance and oil and gas projects and contracts in the decades to come.

Arbitration in the UAE

Although Dubai is located in the United Arab Emirates, it is essentially 'sovereign' with respect to international arbitration. From as far back as 1975, international arbitration has been popular in Dubai.[78] More recently, the Dubai Financial Centre (DIFC) is strategically positioned as one of the first centres to implement the newest arbitration laws (DIFCAL). The Dubai International Financial Centre (DIFC) (established in 2004) has its own jurisdiction and a stand-alone arbitration law (the "DIFC Arbitration Law")'. Awards originating from the DIFC under the DIFC Arbitration Law are enforceable in Dubai without being subject to the power of judicial review by Dubai Courts.

Investment Opportunities in the Kingdom of Saudi Arabia

Recently (in May of 2013), the Kingdom of Saudi Arabia allocated $2 billion for the development of the Saudi judicial system with special reference to the field of international commercial arbitration. This represents one of the leading developments in the International Chamber of Commerce (ICC).[79] In addition, there have been many improvements in the legislative, regulatory and legal environments of business and commercial activities in the Kingdom and this will stimulate and attract domestic and foreign investments.[80] Included with these changes are King Abdullah's project for the development of the judicial system, the new foreign investment law and last year's new arbitration law.[81] This judicial project incorporates updating and streamlining many legal procedures, rehabilitating and training judicial staff and instituting modern technologies all with a view to boost the Saudi judicial system and increase investor confidence.[82]

Arbitration in the Kingdom of Saudi Arabia

Concurrent with increased large scale contracts, of which the above is only one example, is the fact that the Kingdom of Saudi Arabia recently revised its international commercial arbitration law.[83] Undoubtedly, the fact that the arbitration law adopts many UNCITRAL provisions is a signal that with the stable system of arbitration to help mitigate adjudicatory risks, entering into large scale contracts is now an even more favourable option.

The Creation of Free Zones to Attract Investors

The creation of free zones to attract investors is especially notable in the well-known case of Dubai. As part of a complete package, Dubai has created a free zone, together with its own arbitration courts and law. Dubai recently drafted two important International Commercial Arbitration Law instruments which have been extremely important for a number of reasons. They have further added confidence in investor perceptions of Arab Arbitration Centers and Laws. As a result they have increased investment and business in the United Arab Emirates. They have raised the number of commercial contracts entered into by foreign investors as well as commercial parties with each other employing the DIFC[84] and DIAC[85] as Seats of arbitration. As a result of the foregoing, the economy of the UAE as a whole as well as the overall GCC economies have benefited. The benefits have occurred as a result of regional trade agreements, such as the Riyadh Convention, which increased business, greater confidence and is a buffering element against the Global Financial Crisis of 2008, which is arguably a result of the DIFC and DIAC, had less of an impact on Dubai than other regions or states.

This is part of the larger trend in the history of legal developments in foreign investment: 'Many states, particularly the developing ones, adopt special tax regimes to attract foreign investment, usually in the form of tax reduction or exemption. Some reductions and exemptions may apply to all foreign investment but the majority of them are targeted at certain industries or regions.'[86] The Gulf Arab States are certainly not only key players in this positive development but also trend setters.

Investment Opportunities in the Oil and Gas Industry in Iraq

Updates to Iraqi investment laws are worth considering. The case of Iraq is a valuable study of a Gulf Arab state with a complex history because it demonstrates both strengths and weaknesses, as well as opportunities and threats. These terms are part of a widely used investment and management assessment tool, known as a SWOT analysis which helps investors weigh the overall situation of a country to assess political, country, adjudicatory and investment risk. It is the author's view that Iraq is now in a position to offer more strengths and opportunities than ever before. A recent trade delegation at the Australian Arab Chamber of Commerce in Sydney (attended by the author) and hosted the Australian Ambassador to Iraq, the Iraqi Ambassador to Australia and a number of ministers and key focal investment people. Iraq has made significant reforms to create greater transparency, opportunity and ease for investors seeking to enter the Iraqi markets and this ought to be taken very seriously indeed. Overall, in spite of geopolitical and transportation issues which have in the past had an impact on the oil and gas industry in Iraq, there remain tremendous opportunities. Iraq is an excellent study of how investment in a country can support both the industry and the economy.

The overall picture in Iraq is the following. Iraq produces 143 billion barrels in oil reserves, 9% of the global total. Iraq is the world's third largest exporter, after Saudi Arabia and Russia. Oil there is easily accessible and easy to drill. Western Iraq continues to hold more opportunity and possibility as it has not yet been explored or mined. In 2012 figures for Iraq show it produced 3 million barrels a day (b/d) for the first time since 1990. The government estimates an increase to 10 m b/d by 2020.[87] Indeed, Turkey has a $17 billion trade with Iraq.[88]

Iraq, for the purposes of this analysis of the current oil and gas market, is here divided into three key zones of focus. They are Kurdistan, Najf, and Northern Iraq.

Kurdistan in the north has a highly stable infrastructure, with 24 hour electricity. Kurdistan's GDP relies entirely on oil revenues, earning it $8 billion a month.[89] The International Energy Agency suggests that exports could double by 2020[90] and this is certainly feasible in light of the fact that Kurdistan has 23 known oil fields. Kurdistan currently has fifty oil and gas deals with foreign firms and is exporting oil to Turkey in trucks.[91] The ease of transport to nearby locations is certainly a strength. Exxon Mobil has made a production-sharing agreement with the Kurdistan government.[92]

In the south of Iraq, in Najf, the Chamber of Commerce has found that the oil industry supports $ 7 billion in foreign investment.[93] Around approximately 32-36 known oil fields are located near or in this region. Both Najf and Kurdistan have excellent infrastructure and transportation.[94]

In the northern area of Iraq are approximately 13 known oil fields. The World Bank ranking of Iraq puts it at 165 out of 185 in terms of its Ease of Doing Business Index.[95] The reasons given by the World Bank are the following: shipping containers into/out of the region take approximately four times as long with three times the cost, access to credit is harder and the procedure to start a business is relatively slower, electricity (vital to operations) is on only a few hours at a time, and this limits productivity.[96] As of 2007, petroleum laws endorsed by Iraq's cabinet to bring new investment have yet to pass parliament.[97] The author strongly encourages the Iraqi government to continue to make and enter into force reforms and changes that can allow greater investment opportunities. There is currently a need to spend money on pipelines, ports and other infrastructure.[98] It is highly feasible that greater investment can fund these types of internal and external infrastructures which will in turn attract greater investment.

The author submits that notwithstanding the foregoing, the new Iraqi reforms have made a significant positive improvement that is yet to be measured and as investment increases into the country, the infrastructure matters will in turn respond favourably. One of the recent reforms is the provision to allow a greater percentage of foreign ownership. The author is therefore optimistic that these countries analyses throughout this paper, although highly diverse, all have significant strengths to offer to investors who study opportunities carefully and prepare to deal with foreseeable risks.

Kuwait

Kuwait is aiming for renewable energy sources in order to lessen dependence on oil.[99] The goal is to lessen dependence on petrol and gas that fuels electricity by 15 percent by 2013.[100] Kuwait's daily consumption of oil to generate electricity is 350,000 barrels and this costs USD two billion.[101] Kuwait has recently held important workshops that specialise in oil and gas pricing, and the effects of these prices on international oil demands and current world events, according to Dr Sa'ad Salem Al-Jandal, head of the energy program of the Kuwait institute for Scientific Research.[102] Shifting dependency onto renewable energy sources such as solar and wind power has implications on oil prices and oil demands.[103] Pricing is important for setting a scale between different energy sources.[104] A major object of the institute is the study of renewable energy sources in order to present results to decision makers.[105] The institute applies energy policy and then monitors the outcome on the policies of other nations.[106] All Gulf countries are currently aiming to reduce the use of oil in electricity by substituting it for renewable sources.[107] This is particularly relevant for Kuwait due to its lack of oil required for fuelling power plants.[108] The Al-Shegaya renewable energy plant is one such example of using renewable sources.[109] The goal to make generating electricity easier for power plants by reducing dependence on oil, as such large oil consumption is unsustainable.[110]

The author is of the view that sustainable and renewable energy sources not only hold merit for their environmental impact, which is less harmful than other sources of energy, but this policy shift from oil to renewable sources also serves to lessen adjudicatory risk because it means that long-term contracts will not face a situation where an energy source has run out or a power-plant has had to stop operations during the life of the contract. The reasons for these possibilities are many and may include resource nationalism which can include direct and indirect expropriation. Renewable sources of energy such as solar and wind power are less environmentally destructive as well as sustainable and therefore reliable. This contributes to a more stable regulatory framework overall and that translates to lower adjudicatory risk. Further, it is highly commendable that energy policy across the Gulf is taking into consideration environmental factors. Sustainable development and renewable energy sources are good policy. This topic ought to be further explored for both its scientific implications and legal implications, in future research.

Economic Value of Investment[111] Treaties

In the most general sense, the utility of investment treaties is based on the protection they offer to investors vis-a-vis the host state.[112] The standard list of protections (which may be present in whole or in part in most or some BITs) generally include: fair and equitable treatment, full protection and security, umbrella clauses, access to justice, fair procedure, and protection from denial of justice, emergency provisions, necessity provisions and those for armed conflicts, and force majeure, preservation of rights, arbitrary or discriminatory measures, national treatment, most-favoured-nation treatment, and transfer of funds. In summation, 'Investment treaties effectively allow foreign investors to self-insure against sovereign risk.'[113]

Another major benefit of investment treaty protection is concerned with the lending of capital:

> At the planning and financing stages, where an investment is covered by a trade or investment treaty, its sovereign risk is reduced. Because the investment has lower risk, it is easier to sell to banks, and this means that the finance costs of the project will be lower. While the precise finance cost reduction that the investor enjoys will vary depending on a range of factors (including the scale of the project and the sovereign risk profile of the host state involved), having treaty coverage at least puts the investor in a good position to negotiate with its lenders for lower rates.[114]

Bilateral Investment Treaties (BITs)

The benefits of Bilateral Investment Treaties (BITs) are connected to the matter of adjudicatory risk and hence the need for arbitration (which will be discussed subsequently). Bilateral Investment Treaties exist to encourage and protect international trade and commerce. They are essential in protecting the rights of investors in a situation where those investors are contracting with a State party, in order to level the playing field between a State and private commercial actor(s) in terms of sovereign risk. The relationship of BITs with arbitration is made clear here: 'The overarching purpose of this system is the creation of stable investment environments and thus the general promotion of foreign investments in order to foster economic development and human well-being. Arguably, it provides the investor with procedural and substantial rights which are directly enforceable vis-a-vis the host State on the international plane and thus avoids domestic adjudication and diplomatic protection.'[115] The rights provided for in a BIT are substantial.[116]

Investor-State Arbitration

Certain legal provisions apply to investor-state contracts and disputes whether they deal with oil and gas or renewable energy. The fact that most of these contracts fall under investor-State arbitration,[117] either ICSID or ad hoc, means they are therefore governed by a legal framework that must evolve with the commercial needs of both the investor and the contracting government party. There are three important legal doctrines that impact these types of contracts. The first is the doctrine of public policy. The second is the notion of an administrative contract and the third is the need to understand essential elements of international public law which governs treaties and international agreements when state parties are involved, in contrast to international private law which deals exclusively with commercial actors. International public law automatically activates the doctrine of sovereign immunity as an important consideration to the drafters and negotiators of investor-state contracts. The lessons learned from the early oil concessions with their nationalisations, arbitrations, renegotiations have value to future renewable energy contracts. The conclusion from this analysis in light of current economic and energy trends[118] in the industry is that Middle East oil is not dead. These trends have important economic and strategic implications. For example: 'In Iran, Libya and Algeria, the population is addicted to low energy prices and high oil-funded social spending. All of that is in jeopardy as countries struggle to boost production- and new revenue is needed to quell potential unrest or prevent new problems.'[119]

Economic Benefits of Investment Treaties, International Arbitration and Regulatory Framework Reform

The impact of the recent[120] suite of newly proposed legal changes in the United Arab Emirates Commercial Arbitration Laws on the local economies of the GCC States has been highly significant and the effects of it in the future are arguably going to gain in further relevance. The recent Global Financial Crisis of 2008 has demonstrated that the strength of the DIFC[121] and the DIAC[122] the two most recent United Arab Emirates International Commercial Arbitration Centres has been highly substantial.

The economic benefits of these legal reforms has had an impact on the GCC States by way of regional trade agreements, attracting further investment into the UAE and

GCC and in facilitating and enhancing business relationships with the GCC countries as international commercial arbitration arguably is linked to better business dispute resolution, face saving and ongoing business relations and business ties as well as greater trust in the Arab International Commercial Arbitration regional centres. One empirical form of evidence of the increased confidence of the Dubai institutional arbitration centres of investors and commercial parties is the fact that all the top tier so called 'magic circle' law firms are trying to open branch offices in the UAE as a result of the transparency of the newly reformed Dubai Laws.

Adjudicatory Risk

One of the most important matters facing the oil and gas industry legally is the matter of adjudicatory risk. The author's advice to Gulf Cooperation Council states at this time is that the GCC countries need to manage perceived risks. The matter of adjudicatory risk contains both perceived and actual risk but perceived risk needs to be minimised. There are four factors that contribute to perceived risk. They are (1) the history of the early oil concessions, (2) the fact that they are emerging markets, (3) the potential for resource nationalism,[123] (3) political risk where the Arab Spring has become a liability. Steps can be taken to convey to investors and shareholders that these factors will not pose unnecessary risk and this can be done through the language used in drafting multilateral treaties. The Harmonised International Commercial Arbitration Law Code (HICALC) or Uniform Arab Arbitration Law (UAAL) proposed here can help manage the perception of risk and actually reduce real risk.

Hedging Against Adjudicatory Risk

We have seen in the earlier discussion how the global financial crisis, especially the one in Europe has led to greater third party risk, need for potential contract renegotiation as a result of the reduction of capital. The main source of adjudicatory risk arises from the contract itself. Contract risk management affects legal counsel, contract managers and supply chain professionals. These contracts need to cover a wide territory and anticipate foreseeable and unforeseeable contingencies within a range of interconnected business and managerial forums in order to cover large scale projects and their multiple dimensions. The differences in the suite of contracts and agreements to choose from requires industry and expert knowledge of the different types of risk each one poses. Legal counsel must take steps to protect their clients in order to hedge risk, particularly because legal risk insurance is untenable and does not necessarily cover all the potential losses that could occur.[124] A more feasible solution is to combine risk insurance with treaty protection.[125]

The reforms of the United Arab Emirates international commercial arbitration laws, and particularly those of the cutting-edge institutions such as the DIFC[126] and the

DIAC[127] should be emulated by the other GCC States and already Qatar has drafted a new law whilst Saudi Arabia is considering reforms in this area and Bahrain followed with a Regional Arbitration Centre. International arbitration is a legally binding method of dispute resolution that protects a foreign party from the need to appear before a domestic court. It is distinct from 'expert valuation, mediation, conciliation, 'mini-trial".[128]

Need for Arbitration

The need for arbitration is directly proportional to adjudicatory risk. For arbitration to be considered international, it has to fulfil the following conditions:

> 'As to internationality, the definition given in the United Nations (UNCITRAL) Model Law on International Commercial Arbitration may be taken as reflecting a generally accepted view. Article 1(3) defines international arbitration as one where: (a) the parties to an arbitration agreement have, at the time of the conclusion of that agreement, their places of business in different states; or (b) one of the following places is situated outside the state in which the parties have their places of business: (i) the place of arbitration if determined in, or pursuant to, the arbitration agreement; (ii) any place where a substantial part of the obligations of the commercial relationship is to be performed or the place with which the subject matter of the dispute is most closely connected; or (c) the parties have expressly agreed that the subject matter of the arbitration agreement relates to more than one country.'[129]

There are two major legal enforcement frameworks for international arbitration. The first is ICSID (the International Centre for the Settlement of Investment Disputes) and the second enforcing instrument is the New York Convention (the Convention on the Recognition and Enforcement of Foreign Arbitral Awards 1958). The ICSID framework open only to member states and the awards are published publically whereas ad hoc arbitrations under the New York Convention are private and confidential.

Provisions for a Multilateral Investment Treaty (MLT) for Adoption in the MENA (HICALC)/ (UAAL)

In light of the discussion on international investment and international arbitration – including international arbitration provisions in BITs and MLTs, the author argues that in light of the foregoing important considerations, a regional Multilateral Investment Agreement (MIA) in the form of a Harmonised International Commercial Arbitration

Law Code (HICALC) or Uniform Arab Arbitration Law (UAAL) can act as a way to attract further oil concessions that are fair to both parties and that minimises the adjudicatory and legal risks that could foreseeably arise. A regional free trade agreement would not have the same weight as a multilateral investment agreement or a multilateral[130] investment treaty. In fact, a multilateral investment agreement has great potential as: 'The collapse of multilateral investment agreement (MIA) negotiations and the stalemate in the Doha Round have led some countries, such as Japan, to abandon partially its multilateral belief and promote bilateral or regional free trade agreements[131] instead. However, it is believed that an MIA is a desirable outcome and will eventually be achieved.'[132] The climate is ripe for the opportunity to bring Gulf Arab States together in mutual interest and cooperation to conclude a multilateral investment agreement amongst themselves and perhaps that also includes other key trading parties from outside of the region.

Conclusion

Legal provisions and trends concerning oil agreements and concessions are still relevant and perhaps even more so in light of the higher demand for oil as it is predicted to run out during the gap in setting up proper physical and regulatory legal structures to create the necessary infrastructure and government support for renewable energy contracts.

A key element underpinning the economic vitality of Gulf Arab states in the energy sector has to do with the successful forging of foreign partnerships and contract pre-negotiation, negotiation, drafting and subsequent dispute resolution. The matter of dispute resolution has been discussed earlier in this paper in the examination of international arbitration and reforms especially in the United Arab Emirates (UAE) and the Kingdom of Saudi Arabia. In addition to reforms or changes in the international arbitration laws, are the creation of free zones and arbitration courts. In other MENA nations, such as Egypt, similar reforms took place in the recent past including updating international arbitration laws, the creation of free zones and viable investment laws. The matter of contract negotiation requires an understanding of the different natures of the contracts and the risks they may bring. The earlier discussion of the different types of contracts has provided insight into this matter and can provide a useful starting point for considerations for future renewable energy contracts. The matter of relations with foreign parties is inherently connected with all of the above because a solid regulatory framework made up of international arbitration laws, free zones, investment laws and bilateral treaties that support the smooth operation of business and projects in the energy field is necessary. The author submits one final advice in this regard. A regional free trade agreement,[133] or HICALC, if drafted correctly, can further encourage and support both fossil fuel and renewable projects now and in the coming decades. A region free trade agreement can protect the economic interests of the Gulf Arab States by creating cooperation, trust, and stability so that potentially higher-risk projects can be created, maintained and resolved in a competitive and changing energy and economic landscape. The HICALC addresses the major legal and doctrinal matters that can arise in investor-state contracts and disputes.

Appendix 1:
A Harmonised International Commercial Arbitration Law Code (HICALC)

Travaux Preparatoires of the HICALC

These provisions were presented to His Royal Highness Prince Faisal Al-Turki at the Gulf Research Meeting 2011 as recommendations for revising the international arbitration law of the Kingdom of Saudi Arabia. One of the main recommendations was to make the arbitration law res judicata, or binding, which it previously was not. A short time later the Kingdom of Saudi Arabia drafted a new international arbitration law and one of its changes was the reform of the law making it res judicata, or binding. This is noteworthy especially in light of the early oil concession history which was unfavourable toward the Arab parties. This change in the law signals a very positive change, one that is based on renewed trust between Arab and Western European parties and this bodes very well for the investment and legal climate in the GCC countries and especially for the Kingdom of Saudi Arabia in attracting new investment in the energy industry. The major principles serving as the foundations for the code are the universal principles of law discussed in this thesis including the common principles of law found at civil, common and Islamic law. Both threads of principles, i.e., the universal and the ones common to the three traditions, were derived through the comparative analysis the author engaged in as a result of the research. These provisions are a response to the distinctive and special features of the MENA legal climate; its laws, traditions, history and unique concerns.

Matters of *Compétence de la Compétence*, Expropriation, Interest, Public Policy, Sovereign Immunity, Precedent and res judicata are dealt with in the HICALC. For a complete draft of the HICALC please see Ayad, Mary B., A Model Harmonised International Commercial Arbitration Law Code for the Middle East and North Africa. Springer Publishers, 2015. The Article concerning res judicata presented to His Highness is reprinted below:

Article VII of the HICALC *res judicata*

Any Award of an Arbitral Tribunal *shall* be final and binding on the parties. The parties *shall* undertake to comply fully and promptly with any award without delay and *shall* be deemed to have waived their right to any form of recourse insofar as such waiver can validly be made. (Adopted from the IBA *Guidelines for Drafting International Arbitration Clauses.*)

Judicial review of arbitral tribunals *shall* be restricted to determine the public policy considerations of an award *after* a tribunal has given its decision.

Appendix 2

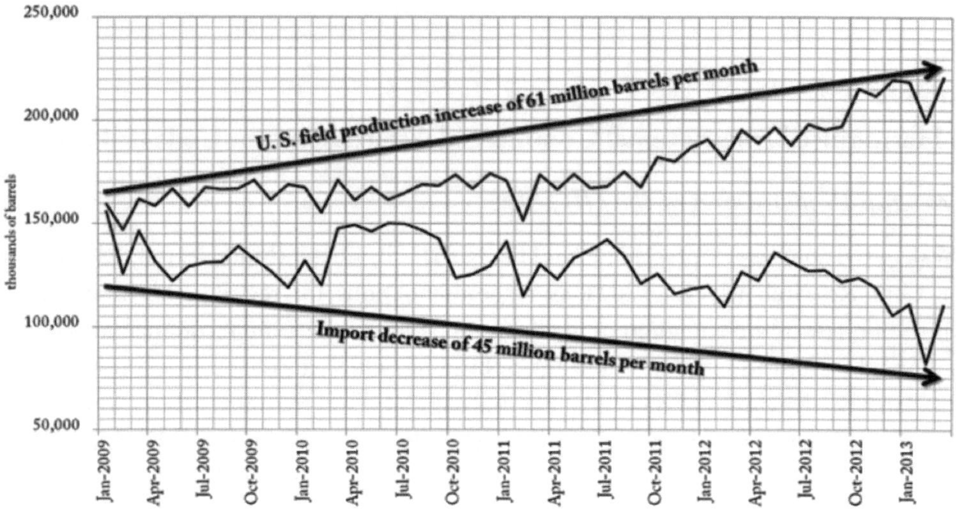

Source: Senate Republican Policy paper, Senator John Barrasso, Chairman, U.S. Shale Oil Revolution "Grave Concern" for OPEC, June 19, 2013.

Appendix 3

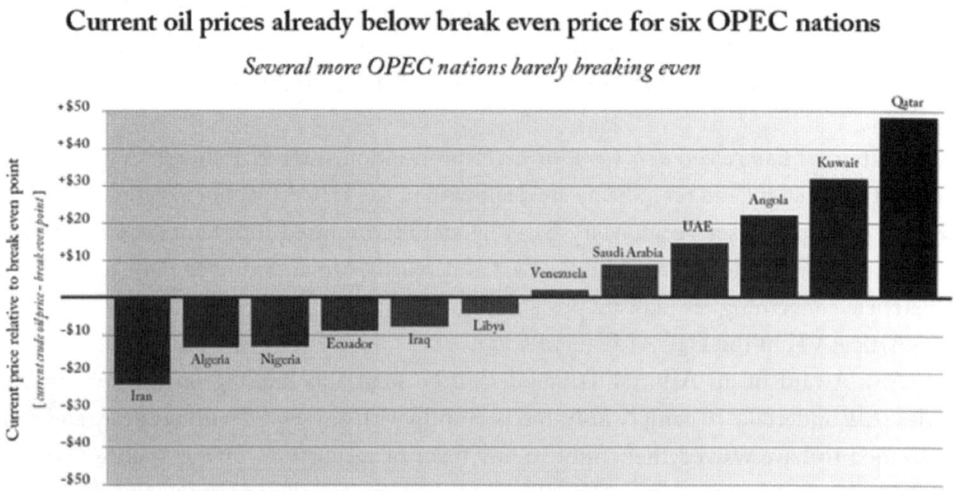

Source: Senate Republican Policy paper, Senator John Barrasso, Chairman, U.S. Shale Oil Revolution "Grave Concern" for OPEC, June 19, 2013. http://www.rpc.senate.gov/policy-papers/us-shale-oil-revolution-grave-concern-for-opec, retrieved 29 May 2014.

Appendix 4:
Relevant Legal Journal Articles, Industry Publications and White Papers

Ayad, MB, The Doctrines of Public Policy and Competence in Investor–State Arbitration, **Arab Law Quarterly**, Volume 27.4, 2013.

Ayad, MB, An amicus curiae Submission to the Brazilian Government in the Matter of Introducing Bilateral Investment Treaties (BITs), **Transnational Dispute Management (TDM)** 14 June 2013.

Ayad, MB, 'Harmonization of Custom, General Principles of Law, and Islamic Law in Oil Concessions' (2012) 29 Journal of International Arbitration, Issue 5, pp. 477–518.

Ayad, MB, The Vienna Convention as Authority for the use of Precedent as Customary Practice in International Arbitrations of Oil Concessions and Investment Disputes in North Africa and the Gulf Arab States; or a lex mercatoria for a lex petrolea, **Journal of World Investment and Trade**, Volume 15 (2013) forthcoming.

Ayad, MB, A Theory of International Commercial Arbitration and International Investment Arbitration Law and Practice For a Uniform Arab Arbitration Law, **Vindobona Journal of International Commercial Law and Arbitration** (2012) 16 VJ 113-122.

Ayad MB, International Commercial Arbitration and Harmonization of Contract Law with a Focus on Reform in the Mena Region, **Vindobona Journal of Comparative Law and Arbitration** 12.2 (2008): 169-174.

Ayad MB, Harmonization of International Commercial Arbitration Law and Sharia: The case of pacta sunt servanda v. ordre public; The use of ijtihad to achieve higher award enforcement, **Macquarie University Journal of Business Law**, 6, (2009).

Ayad MB, Towards a truly harmonized international commercial and investment arbitration law code (HICIALC): enforcing MENA- Foreign Investor arbitrations via a single regulatory framework: A new map for a new landscape. **Macquarie Journal of Business Law** (2010) Vol 7 285-322.

Ayad MB, International Commercial Arbitration Award Enforcement at the Crossroads of Sharia Law and Ordre Public in the MENA: Paving the Golden Path towards Harmonization, **Journal of World Investment and Trade** 10.5 (2009).

Ayad MB, Investor risks due to 'Sovereign Immunity' pleas in Court Rulings on Arbitral Award Enforcement of MENA-FI Investments can be mitigated via a Harmonized international commercial arbitration law code, **Journal of World Investment and Trade** Volume 11 No.5 (2010), pp. 753-787.

Ayad MB, Harmonized Commercial Law Principles in Investor–State international commercial arbitrations amongst Euro-Med Partners, **Id-Dritt Law Journal**- Maltese Law Student's Society XXI (2011).

Ayad MB, The Quest of the Holy Grail: A Harmonized international commercial arbitration law code (HICALC) in Investor–State Arbitrations of Oil Concessions and Foreign Investment Treaties and Agreements, 2011 **Madrid International Commercial Arbitration Centre Journal.** Centro Internacional de Arbitraje, Mediacion y Negociacion Instituto Universario de Estudios Europeos, Universidad San Pablo Madrid, Arbitraje, Revista de Arbitraje commercial y de inversions Volumen IV 2011 March/April, pp. 335-381, March 20, 2011.

Ayad MB, A Harmonized International Commercial and investment Arbitration Law Code (HICIALC) Regulating Judicial Review of Investor–State Arbitration Disputes Can Uphold res judicata of Arbitration Awards and Tribunals from Undue Appeals and Annulment due to State Intervention, Vol XV **International Trade and Business Law Review** (2011).

Ayad MB, The Legal and Economic Implications of World Trade Organisation Ascension By the Gulf Arab States, **Transnational Dispute Management** 3 (2011).

Ayad MB, The Quest for the Holy Grail of Social Justice: Substantive & Procedural Law Provisions for Amicus Curiae in Investor–State ICSID Arbitration Law & Practise, **Macquarie Journal of Business Law** (2012) Vol 9.

Ayad MB, Australia's Bilateral Investment Treaty (BIT) Stance- an open amicus curiae submission, **Macquarie Journal of Business Law** (2012) Vol 9.

Ayad MB, Current Debates in Refugee Law: Analysing a Gender Based Claim: The Impact of A New Interpretation of Refugee Law and the Canadian Guidelines, **University of Malta Mediterranean Journal of Human Rights**, December 5, 2011.

Ayad MB, United Nations Treaty Body Reform; Human Rights are not Optional Edit **University of Malta Mediterranean Journal of Human Rights**, Volume 13 2009 (2009).

Ayad MB, The Office of the Ombudsman: A Vehicle for Human Rights Protection via Good Governance and Rule of Law in Democratic States, Quis custodiet ipsos custodes? **Universita Degli Studi Kore Di Enna Journal**, Italy (2009).

Ayad MB, Integration is A Human Right, University of Malta Mediterranean Journal of Human Rights. (2008).

Ayad MB, The Regulatory Framework of Egyptian Non Government Organisations (NGO) Management: A Proposal for Legal Reforms in Light of the Arab Spring. Common Ground Publishing, **The International Journal of Knowledge, Culture and Change Management**, 2012.

Ayad MB, Do Rights of Foreign Investors Fall Within The Scope of International Human Rights Law? **Id-Dritt Law Journal XXIII**, March 2013.

Ayad MB, What Now for Egypt? **Global Arbitration Review, (GAR)**, World Edition, March 2nd, 2011, Law Business Research Ltd, London.

Ayad MB, Global View, The Future of the New Egypt, **Australian Centre for International Commercial Arbitration (ACICA) Newsletter**, Quarterly Update, March 2011 Vol 2 No 1 ISSN 1837-8994.

Ayad MB, Investor Report: Egypt News, **Australian Centre for International Commercial Arbitration (ACICA) Newsletter**, Quarterly Update, December 2011 Vol 3 No 4 ISSN 1837-8994.

Ayad MB and Stephen J. Keim SC, Political Speech Blooms. High Court breathes fresh life into implied right of speech, Australia: **Justinian**, 2011.

Ayad, MB and Stephen J. Keim, SC, Morsi Crosses a Great River, Australia, **Justinian**, Dec. 15, 2012

Ayad MB and Stephen J. Keim SC, Anti-Mubarak protesters have Egyptian law on their side, **Courier Newspaper**, Queensland, Australia Feb. 2011: 68-68.

Ayad MB, High Court Irons Out Cultural Differences, **Online Opinion** Wed 10 Oct 2012.

Ayad MB, Egypt at a Crossroads, **Online Opinion**, 12/12/2012.

Ayad MB, A Model Harmonised International Arbitration Law Code for the Middle East and North Africa, **Springer Publishers**, September 2013, Forthcoming. ISBN 978-3-642-38194-2.

Notes

1 This paper is a revised and expanded version of the author's doctoral dissertation that was presented at the Gulf Cooperation Council (GCC) Gulf Research Meeting at Cambridge University in 2011, 2013 and 2014. This draft is specially prepared for the Gulf Research Meeting 2014 Cambridge University, UK, Workshop Eleven: "The Changing Energy Landscape in the Gulf: Strategic Implications."

2 McKinney's Consolidated Laws of New York, Energy Law, Thomson West, NY, 2004 at p. 6, '"Petroleum products" shall include all products refined or rerefined from synthetic or crude oil or oil extracted from other sources, including natural gas liquids.'

3 International Chamber of Commerce Court of Arbitration Rules.

4 London Court of International Arbitration Rules.

5 Hong Kong International Arbitration Centre Administered Rules.

6 Dubai International Arbitration Centre.

7 Using Investment Treaties to Manage the Risk of Resource Nationalism in the Oil Industry, Oil Council Expert Insight, Paul Sandosham, Sam Luttrell. 'One of the great advantages of the ISA system is that a successful claimant does not need to enforce the award in territory of the host state. The enforcement treaties allow an award creditor to enforce against the state debtor anywhere it has assets, subject to the ability of the award-creditor to enforce against the state debtor anywhere it has assets, subject to the ability of the award-creditor to overcome any defences of sovereign immunity that the state may raise

(in accordance with the local law of the state in which enforcement is sought). This strategy of pursing enforcement against host state assets in other countries is sometimes referred to as "Hot Oil Litigation", an expression that comes from the satellite actions brought by oil companies following the expropriation of their assets in Libya in the 1970s.'

8 Judicial review is automatic in the UAE. Court intervention is guaranteed. Dubai however, has its own arbitration courts outside of the UAE jurisdiction.

9 M.N.H. Comsan, Nuclear Energy Perspectives in MENA Countries, 2nd International Conference on Energy Systems and Technologies, 18-21 February. 2013, Cairo, Egypt, at p. 9, 'World demand for energy is set to increase significantly in the coming decades, spurred by economic and demographic growth, especially in developing countries. Unless, current policies are changed, this demand for energy will be met mainly by burning fossil fuel, at the cost of escalating emissions of carbon dioxide and the associated risk of global warming.'
The environmental is only one concern. Economic realities, such as the costs of consuming more nonrenewable sources than are exported is another valid and substantial concern.

10 An example of a recent risk in renewable energy is the ICSID (International Centre for the Settlement of International Disputes) action against Spain in the solar panel contract. PV Tech, Connor Ryan, March 14, 2014, Friday Focus: Masdar lawsuit may turn tide against Spanish solar cuts. http://www.pv-tech.org/friday_focus/masdar_lawsuit_may_put_pv_investors_over_the_top_in_battle_against_spanish, retrieved on 29 May 2014:
'In 2008 alone, Spain was responsible for half of the world's solar power facilities in terms of wattage. Unfortunately, the financial downturn in late 2008 had major repercussions for Spain, as its solar industry plummeted. Madrid would soon face a growing deficit in its energy budget of around €4.5 billion (US$6.1 billion) by 2012. As a result, Spain, once the crown jewel of the solar power industry, gutted its market by removing its FiTs, cutting subsidies and capping the rate of return for investors. Needless to say, these measures have incensed the thousands of PV investors who will now see little benefit in their ventures in Spain. What has followed has been a seemingly endless carousel of lawsuits directed towards the Spanish government, as PV investment, development and installation companies such as Foresight Group and European Energy have lined up to launch grievances again Madrid's reforms and subsequent cuts. Yet another major PV player entered into the fray in February 2014, as a subsidiary of UAE-backed company Masdar filed a legal complaint against Spain before the International Centre for Settlement of Investment Disputes (ICSID) in order to protect its investments from these reforms. Currently, the ICSID has five cases against Spain on record, four of which revolve around renewable energy.'

11 However, it is the author's submission that the matter of environmental policy in the energy industry in the Gulf Cooperation Council countries is linked to sustainable development and therefore this means less adjudicatory risk for large-scale contracts once the infrastructure is in place.

12 McKinney's Consolidated Laws of New York, Energy Law, Thomson West, NY, 2004 at p. 6, '"renewable energy resources" shall include sources which are capable of being continuously restored by natural or other means or so large as to be useable for centuries without significant depletion and include but are not limited to solar, wind, plant and forest products, wastes, tidal, hydro, geothermal, deuterium, and hydrogen.'

13 M.N.H. Comsan, Nuclear Energy Perspectives in MENA Countries, 2nd International Conference on Energy Systems and Technologies, 18-21 February. 2013, Cairo, Egypt, at p. 10.

14 Ibid at p. 12.

15 M.N.H. Comsan, Nuclear Energy Perspectives in MENA Countries, 2nd International Conference on Energy Systems and Technologies, 18-21 February. 2013, Cairo, Egypt, at p. 9, 'The expanding market for nuclear equipment and services in support of the construction of new Nuclear Power Plants (NPPs) in the Middle East and North Africa (MENA) region is growing. To meet increasing energy demands, the last decade has seen countries in the MENA turn to alternative energy sources such as nuclear power and renewable energy. The interest in nuclear energy has intensified in the last few years despite issues raised following the accident at the Fukushima Daiichi NPP in Japan. There are a number of key challenges facing countries in the MENA region that are driving plans towards nuclear power

generation. In particular countries in the region face: increasing energy demands due to population and economy growth, energy independence, reduction of reliance on fossil fuels, increasing revenue on fossil fuel exports and reduction of carbon emissions. As a direct result of these challenges there are significant plans to deliver considerable nuclear generation capacity in the region by 2030.'

16 M.N.H. Comsan, Nuclear Energy Perspectives in MENA Countries, 2nd International Conference on Energy Systems and Technologies, 18-21 February. 2013, Cairo, Egypt, at p. 9, 'Reinforcing nuclear safety and transparency along with setting up and maintenance of efficient governance will obviously strengthen public acceptance of nuclear power. Clearly nuclear energy will continue to have its share in the future energy mix as energy policies of many countries already show.'

The author is of the view that any improvements in extant regulatory frameworks involving pre-existing energy projects will support future renewable energy projects and agreements.

17 The US Shale Oil Boom: Potential Impacts and Vulnerabilities of an Unconventional Energy Source. Policy Brief, Belfer Center for Science and International Affairs, Harvard Kennedy School, Leonardo Maugeri, June 2013.

18 Ibid.

19 Energy Law, S 1-03, para b, sec 76, Article 1: '"Energy resources" shall mean any force or material which yields or has the potential to yield energy, including but not limited to electrical, fossil, geothermal, wind, hydro, solid waste, tidal, wood, solar and nuclear sources.'

20 Energy Law, S 1-03, para b, sec 76, Article 1.

21 Energy Law, S 3-101, State Energy Policy: 'It shall be the energy policy of the state: 1. To obtain and maintain an adequate and continuous supply of safe, dependable and economic energy for the people of the state and to accelerate development and use within the state of renewable energy sources, all in order to promote the state's economic growth, to create employment within the state, to protect its environmental values, to husband its resources for future generations, and to promote the health and welfare of its people; 2. To encourage conservation of energy in the construction and operation of new commercial, industrial, and residential structures....5. to foster, encourage and promote the prudent development and wise use of all indigenous state energy resources including but not limited to, on-shore oil and natural gas, off-shore oil and natural gas, natural gas from Devonian shale formations, small head hydro, wood, solar, wind, solid waste, energy from biomass, fuel cells and co-generation; and 6. To encourage a new ethic among its citizens to conserve rather than waste precious fuels; and to foster public and private initiative to achieve these ends at the state and local levels.'

22 Energy Law, S 10-101, Article 10, at p. 55: 'The legislature herby finds and declares: that adequate supplies of fuel are essential to the health, welfare and safety of the people of the state of New York; that any severe disruption in fuel supplied for use within the state could cause grave hardship and pose a threat to the health and economic well-being of the people of the state; that the federal government has enacted into law the Emergency Petroleum Allocation Act of 1973, as amended by the Energy Policy and Conservation Act of 1975, pursuant to which the state of New York has established a set-aside program...' This is another description of the public interest.

23 Ernst and Young, Supplemental industry sector considerations, oil and gas, 2012.

24 Ibid.

25 Ibid.

26 Ibid.

27 Ibid.

28 Ibid.

29 Ibid.

30 Senate Republican Policy paper, Senator John Barrasso, Chairman, U.S. Shale Oil Revolution "Grave Concern" for OPEC, June 19, 2013. http://www.rpc.senate.gov/policy-papers/us-shale-oil-revolution-grave-concern-for-opec, Retrieved 29 May 2014. 'OPEC members are most troubled by the downward pressure on oil prices from booming North American supply and softening global demand. The Wall Street Journal has summarized the problem: Each OPEC member has a "break even" price at which oil must sell for that country to balance its budget. In the past decades, these

break-even prices rose rapidly as OPEC countries directed oil revenues towards massive investments in infrastructure projects and, more recently in expanding social programs meant to stave off upheaval cause by the Arab revolutions around them. The higher spending of OPEC countries has left them vulnerable to a drop in prices.'

31 The U.S. Shale Oil Boom: Potential Impacts and Vulnerabilities of an Unconventional Energy Source, Leonardo Maugeri, Policy Brief, Belfer Center for Science and International Affairs, Harvard Kennedy School, June 2013.

32 The U.S. Shale Oil Boom: Potential Impacts and Vulnerabilities of an Unconventional Energy Source, Leonardo Maugeri, Policy Brief, Belfer Center for Science and International Affairs, Harvard Kennedy School, June 2013.

33 The U.S. Shale Oil Boom: Potential Impacts and Vulnerabilities of an Unconventional Energy Source, Leonardo Maugeri, Policy Brief, Belfer Center for Science and International Affairs, Harvard Kennedy School, June 2013.

34 Using Investment Treaties to Manage the Risk of Resource Nationalism in the Oil Industry, Oil Council Expert Insight, Paul Sandosham, Sam Luttrell.

35 Zouhair A. Konfol, Protection of foreign investment, A.W. Sijthoff, Leiden, 1972, at p. 43.

36 This principle has its support at both international public law, French public policy and Islamic jurisprudence.

37 Jenik Radon, The ABCs of Petroleum Contracts: License-Concession Agreements, Joint Ventures and Production Sharing Agreements, Chapter Five, Covering Oil. A Reporter's Guide to Energy and Development Edited by Svetlana Tsalik and Anya Schiffrin Revenue Watch Open Society Institute Initiative for Policy Dialogue, New York, 2005, at p. 69. 'The complexity of a PSA depends on the soundness of the legal infrastructure of a state. For example, if a country does not possess basic rules governing petroleum operations, the issues normally covered by such a law will have to be addressed in the PSA. In Short, the less reliable and/or predictable a state's legal system the more issues must be covered and specified within a PSA.'

38 Ibid, at p. 69.

39 Ibid, p. 68.

40 Ibid, at p. 68.

41 Jenik Radon, The ABCs of Petroleum Contracts: License-Concession Agreements, Joint Ventures and Production Sharing Agreements, Chapter Five, Covering Oil. A Reporter's Guide to Energy and Development Edited by Svetlana Tsalik and Anya Schiffrin Revenue Watch Open Society Institute Initiative for Policy Dialogue, New York, 2005, at p. 69. 'All financial and operational risk rests with the international oil companies. The host government does not risk losses other than the cost of the negotiations (mainly fees paid to advisers). At most, the host government loses an opportunity but suffers no material loss if an exploration or development project fails. Should a project not be pursued in accordance with the terms of an exploration or development program, a government can still, if the PSA is drafted well, cancel or terminate the deal or bring in another oil company. A host government has the added advantage that it shares any potential profits without having to make an investment unless it agreed to do so.'

42 Jenik Radon, The ABCs of Petroleum Contracts: License-Concession Agreements, Joint Ventures and Production Sharing Agreements, Chapter Five, Covering Oil. A Reporter's Guide to Energy and Development Edited by Svetlana Tsalik and Anya Schiffrin Revenue Watch Open Society Institute Initiative for Policy Dialogue, New York, 2005, at p. 70. 'A contractual provision can be more easily contested, and even violated, than a statute or regulation.'

43 Using Investment Treaties to Manage the Risk of Resource Nationalism in the Oil Industry, Oil Council Expert Insight, Paul Sandosham, Sam Luttrell.

44 Using Investment Treaties to Manage the Risk of Resource Nationalism in the Oil Industry, Oil Council Expert Insight, Paul Sandosham, Sam Luttrell.

45 Ibid, at p. 65.

46 Ibid, at p. 65.

47 Ibid, at p.65.

48 Jenik Radon, The ABCs of Petroleum Contracts: License-Concession Agreements, Joint Ventures and Production Sharing Agreements, Chapter Five, Covering Oil. A Reporter's Guide to Energy and Development Edited by Svetlana Tsalik and Anya Schiffrin Revenue Watch Open Society Institute Initiative for Policy Dialogue, New York, 2005, at p. 66. 'It is the nature of the JV that the list of issues to resolve is long. Because a JV demands that the parties do things jointly, by not resolving material issues prior to entering into a JV, the parties only postpone a potential disagreement or a stalemate, if a JV is a 50-50 deal. JVs require painstaking negotiations over an extended period of time to ensure that all matters are thoughtfully addressed and that the parties agree on how to work with each other.'

49 Ibid, at p. 66.

50 Ibid, at p. 67.

51 Ibid, at p. 67.

52 Ibid, at p. 63.

53 Ibid, at p. 63.

54 Ibid, at p. 63.

55 Ibid, at p. 63.

56 Ibid, at p. 63.

57 Ibid, at p. 63.

58 Ibid, at p. 65.

59 Mid East Online, UAE leads MENA region in value of energy contracts, 21/07/2013, http:// www.Middle-east-online.com/english/?id=60254, retrieved on 16/05/2014. 'The Upper Zakum Field Development Project has a budget of 15.6 billion.'

60 Ibid.

61 Ibid.

62 Ibid.

63 Mid East Online, UAE leads MENA region in value of energy contracts, 21/07/2013, http:// www.Middle-east-online.com/english/?id=60254, retrieved on 16/05/2014. Other similar contracts to the Zakum contract are those of 'the Umm al-Lulu projects for offshore oil field development totaling a further $2 billion to the total value of investments in the oil and gas sector in the Emirates.'

64 Mid East Online, UAE leads MENA region in value of energy contracts, 21/07/2013, http:// www.Middle-east-online.com/english/?id=60254, retrieved on 16/05/2014. Additionally, 'Qatar Petroleum (QP) awarded Taiwan's CTCI and Japan's Chiyoda Corporation with a $1.2 billion contract to expand its Ras Laffan Refinery.'

65 Ibid.

66 Ibid.

67 Ibid.

68 Duration is a major condition for a contract to classify as an investment contract. Please refer to the Salini test in the section on international investment for the legal definition. Also, Using Investment Treaties to Manage the Risk of Resource Nationalism in the Oil Industry, Oil Council Expert Insight, Paul Sandosham, Sam Luttrell. 'The definition of investment is usually broad and based on a list of examples, such as moveable and immovable property, shares and stock and other forms of interest in a company, rightful claims to money or to any performance under a contract having financial value, intellectual property rights and business concessions (such as licenses and permits). To use the example of the ACIA, in this multilateral treaty the term 'investment' is defined to include: claims to money or to any contractual performance related to a business (ACIA Article 4(c)(iv); rights under contracts, including Turnkey, construction, management, production or revenue-sharing contracts (ACIA 4(c)(v)); and business concessions required to conduct economic activities and having financial value conferred by law or under a contract, including any concessions to search, cultivate, extract or exploit natural resources (ACIA Article 4(c)(vi)).

69 Scholars agree that there were disastrous consequences for Arab parties.

70 Rodolf Dolzer, Christian Schreuer, Principles of International Investment Law, Oxford University Press, Oxford, 2008, at p. 72.
71 Professor Sornarajah has written about the early history of oil concessions in The International Law on Foreign Investment, Cambridge University Press, 2010, at. P. 38.
72 Rodolf Dolzer, Christian Schreuer, Principles of International Investment Law, Oxford University Press, Oxford, 2008, at p. 73.
73 Rodolf Dolzer, Christian Schreuer, Principles of International Investment Law, Oxford University Press, Oxford, 2008, at p. 73.
74 LLuis Paradel, Andrew Newcombe, Law and practice, standards of treatment of investment treaties, Kluwer Law International 2009, p. 19 'In a series of cases, newly independent and developing States, asserted that, upon independence, States were entitled to review concession agreements that had been granted by colonial power, and furthermore, maintenance that compensation for the expropriation of property would be based on national laws.' This is clearly a response to colonialism.
75 Rodolf Dolzer, Christian Schreuer, Principles of International Investment Law, Oxford University Press, Oxford, 2008, at p. 73.
76 There is potential scope that this railway project will expand throughout the GCC and allow infrastructure development that can one day link the GCC countries. This is a valuable investment in regional infrastructure.
77 Quran, Surah an-Nisa, verse 34.
78 Article 8 of the Dubai Chamber of Commerce and Industry Law No. 2 of 1975 and its amendments established the rules of Commercial Conciliation and Arbitration under the 'Centre for Commercial Conciliation and Arbitration' of Dubai.
79 Abdul Hannan Tago, Saudi Arabia to develop commercial arbitration, Arab News, 30 May 2013, www.arabnews.com. This was decided at the Council of Saudi Chambers consultation meeting for ICC's World Trade Agenda and G20s advisory group. Officials discussed the involvement of global trade in forming world economic policies through active organizations like the G20 and the World Trade Organization (WTO). Saudi Arabia is ranked 12th in the 'Ease of Doing Business Index'.
80 Abdul Hannan Tago, Saudi Arabia to develop commercial arbitration, Arab News, 30 May 2013, www.arabnews.com.
81 Ibid.
82 Ibid.
83 The recent reforms to Saudi Arabia's international arbitration law have included a new provision making international arbitration with the Kingdom legally binding, or *res judicata*.
84 Dubai International Financial Centre.
85 Dubai International Arbitration Centre.
86 Wenhua Shan, (gen ed), The legal protection of foreign investment, a comparative study, Hart Publishing, Oxford and Portland, Oregon, 2012, at p. 43.
87 Briefing Iraq Ten Years On, The Slow Road Back, The Economist, March 2nd, 2013, p. 23.
88 Briefing Iraq Ten Years On, The Slow Road Back, The Economist, March 2nd, 2013, p. 26.
89 Ibid.
90 Ibid.
91 Ibid.
92 Ibid.
93 Ibid.
94 Ibid.
95 Ibid.
96 Briefing Iraq Ten Years On, The Slow Road Back, The Economist, March 2nd, 2013, p. 23.
97 Ibid.
98 Ibid.
99 Kuwait News Agency, 13 June 2013.
100 Ibid.

101 Ibid.
102 Kuwait News Agency, 13 June 2013.
103 Ibid.
104 Ibid.
105 Ibid.
106 Ibid.
107 Ibid.
108 Ibid.
109 Ibid.
110 Ibid.
111 The definition of an investment can be found in the Salini Test. Salini Costruttori S.p.A. and Italstrade S.p.A. v. Kingdom of Morocco, ICSID Case No. ARB/00/4, Decision on Jurisdiction of 23 July 2001, 42 ILM, 2003 [JTB 2], at ¶ 30. The ICSID Convention does not define "investment". ICSID precedents have relied on the Salini test which identifies the following elements as indicative of an "investment" for purposes of the ICSID Convention: (i) a contribution, (ii) a certain duration over which the project is implemented, (iii) a sharing of operational risks, (iv) a contribution to the host State's development.
112 Using Investment Treaties to Manage the Risk of Resource Nationalism in the Oil Industry, Oil Council Expert Insight, Paul Sandosham, Sam Luttrell. 'It used to be that, when their investments were nationalised or expropriated by host governments, oil companies had to turn to their home state governments for protection. This practice, known as "diplomatic protection", was not always available, let alone effective. Ultimately, due to the need to factor in sovereign risk premiums, the costs of doing business abroad became prohibitive for most investors. Outside colonial trade structures, national markets remained largely isolated from one another. However, in the second half of the twentieth century, the idea began to circulate that a system was needed to promote broader foreign investment flows and allow investors themselves to enforce their rights directly against their host states through binding arbitration. States began to conclude investment promotion and protection agreements – what we now call "investment treaties" – that gave their nationals these rights. The fundamental intention of these treaties was to reduce the risks of foreign investment, thereby lowering its costs.
113 Using Investment Treaties to Manage the Risk of Resource Nationalism in the Oil Industry, Oil Council Expert Insight, Paul Sandosham, Sam Luttrell.
114 Using Investment Treaties to Manage the Risk of Resource Nationalism in the Oil Industry, Oil Council Expert Insight, Paul Sandosham, Sam Luttrell.
115 Jan Ole Voss, The impact of investment treaties on contracts between host States and foreign investors, Martinus Nijhoff Publishers, Leiden, 2011, at p. 5.
116 Using Investment Treaties to Manage the Risk of Resource Nationalism in the Oil Industry, Oil Council Expert Insight, Paul Sandosham, Sam Luttrell. They include rules as: (a) Neither contracting state will expropriate, nationalise, or take over, investments made by nationals of the other contracting state in its territory, unless it in accordance with due process of law, for a public purpose, and with the payment of adequate compensation. (b) Each contracting state will treat investors and investments as it would treat its own nationals (national treatment standard). (c) Each contracting state will extend to its investors from the other contracting state any greater benefit that it grants to an investor from a third country, the most favoured nation rule. (d) Each contracting state will accord investors and investments full protection and security, and fair and equitable treatment. (e) In the event of a dispute, the states will submit to binding arbitration in accordance with an agreed upon set of rules.
117 Using Investment Treaties to Manage the Risk of Resource Nationalism in the Oil Industry, Oil Council Expert Insight, Paul Sandosham, Sam Luttrell. 'Almost all investment treaties contain dispute resolution clauses, and most of these allow the parties or their citizens (corporate or natural) to commence arbitration in the event of a breach of their treaty rights. These Investor-State Arbitration (ISA) provisions are critical because they allow the investor to enforce its treaty rights directly against its host state, in a neutral forum that the state does not control, without the need for diplomatic protection. In other words, ISA provisions are what give investment treaties teeth.'

118 Journal Report: Energy. The Wall Street Journal, Monday May 19, 2014. Welcome Back, Big Oil, Benoit Faucon: 'Mexico, Iran and other oil producing nations are trying to entice multinationals back to their countries. The ramifications- for the countries, the companies and geopolitics – could be huge... Algeria is seeking outside help to unlock unconventional reserves. Libya has Africa's largest oil reserves but ranks lower in output and production capacity... For years, Mexico, Iran, Iraq, Algeria and Libya- most of them among the top 10 producers world-wide- were fiercely nationalistic when it came to oil. They either offered Western companies punitive terms to develop their reserves or didn't do business with them at all, controlling their supplies tightly with state-owned companies. Now facing a range of problems, these nations are looking to tap more of their reserves- and they're offering Western companies generous deals to win their help.'

119 Journal Report: Energy. The Wall Street Journal, Monday May 19, 2014. Welcome Back, Big Oil, Benoit Faucon.

120 In 2012 the latest draft of the UAE Federal Arbitration Law was released. The change allows it to comply with the

121 Dubai International Financial Centre.

122 Dubai International Arbitration Centre.

123 Using Investment Treaties to Manage the Risk of Resource Nationalism in the Oil Industry, Oil Council Expert Insight, Paul Sandosham, Sam Luttrell. Here is a view as to perceptions on the part of investors that can lead to a perception of higher risk: 'Related to the political signal, certain characteristic of the society of the host state may also predispose it to resource nationalism. Where the country in question has a young, fast-growing population, the government's need for money to pay for public services and infrastructure may drive it to demand the renegotiation of the terms on which major resource assets are exploited by foreign investors. If these renegotiations fail to achieve the result the treasury seeks, an expropriation may occur. In purely economic terms, the more dependent the host state's economy is on natural resources, the more likely it is that foreign investors in that state will be susceptible to resource nationalism. There are many examples of countries where royalties earned on the exploitation of energy and resources assets represent the vast majority of the government's tax base. Investors need to be aware of the inherent risk of doing business in a state that is overly dependent upon its natural resources, and take steps to ensure that enforceable international recourse is available in the event adverse changes in law or policy occur.'

124 Using Investment Treaties to Manage the Risk of Resource Nationalism in the Oil Industry, Oil Council Expert Insight, Paul Sandosham, Sam Luttrell. 'Investment treaties effectively allow foreign investors to self-insure against sovereign risk. For large-scale, long-term investors like oil companies this is especially significant: the bigger and longer the project, the more expensive it is to procure sovereign risk insurance. Oil companies therefore have a special interest in securing investment treaty coverage for their foreign investments, especially for "too big to insure" projects in developing countries where the risk of resource nationalism is high.'

125 Using Investment Treaties to Manage the Risk of Resource Nationalism in the Oil Industry, Oil Council Expert Insight, Paul Sandosham, Sam Luttrell. '[w]here the investment is treaty-covered, the need to take out sovereign risk insurance is reduced. The investor may, for example, insure the project for a limited period and up to a limited value and, with the comfort of enforceable treaty protection, manage the balance of the risk itself. In so doing, the costs of insuring the investment are reduced, and sometimes eliminated outright (although the ideal combination is to have both sovereign risk cover and treaty protection).'

126 Dubai International Financial Centre.

127 Dubai International Arbitration Centre.

128 Georgios Petrochilos, Procedural law in international arbitration, Oxford, 2004, at p. 3.

129 Ibid.

130 Using Investment Treaties to Manage the Risk of Resource Nationalism in the Oil Industry, Oil Council Expert Insight, Paul Sandosham, Sam Luttrell. '[t]here are a growing number of multilateral investment treaties, such as the Energy Charter Treaty and the ASEAN Comprehensive Agreement (ACIA).'

131 Using Investment Treaties to Manage the Risk of Resource Nationalism in the Oil Industry, Oil Council Expert Insight, Paul Sandosham, Sam Luttrell. 'In terms of their substantive provisions, FTAs tend to e more balanced and complicated than most BITs. For instance, it is not uncommon for FTAs to contain special rules for the qualification of a state measure as an act of expropriation. The ASEAN-Australian-New Zealand Free Trade Agreement (AANZFTA) is one such treaty. The investment chapter of AANZFTA contains an Annex on Expropriation and Compensation. Paragraph 2(b) of this annex clarifies that the expropriation provision includes situations "where an action or series of related actions by a Party has an effect equivalent to direct expropriation without formal transfer of title or outright seizure. This is indirect expropriation.'

132 Wenhua Shan, (gen ed), The legal protection of foreign investment, a comparative study, Hart Publishing, Oxford and Portland, Oregon, 2012, at p. 76.

133 As noted above, a regional Multilateral Investment Agreement (MIA) in the form of a harmonised international commercial arbitration law code (HICALC) or uniform Arab arbitration law (UAAL) can be drafted regionally as an international arbitration law, or a regionally applicable set of rules, or as a treaty depending on the needs of the states. The author hopes that this research can be the beginning stages of the travaux preparatoires.

8

The Potential Role of the Gulf Countries on Re-Shaping the Global Renewable Energy Landscape

Hanan Albuflasa

Introduction

In the next 30 years, the major share of the global energy demand will be shifted toward the developing countries, mainly, China, India and Middle East. The decision that those countries will make, either in term of policies to regulate the energy investment or decisions on the type of energy resources used to supply the growth in their energy need, will impact the global energy landscape in the coming years. At the same time, renewable energy share of the total energy is in rise, which has been projected to reach 31% in 2035[1], with a global investment reached a total of $269 billion[2] in 2012.

China, south America, India is entering the renewable energy sector with both large investments and installations, while the Middle East still progressing very slow, and in some countries it is almost stagnant.

Despite the environmental attractiveness of renewable energy, it will be difficult to compete economically with the dominate energy fuel in the Gulf Cooperation Council (GCC) Countries, natural gas. Hence the governments should take serious incentives to encourage investment in renewable energy, through embarking strong and stable policies to attract investors and major renewable energy players to the region.

This chapter discuss two scenarios for "Renewable Energy Policies" in the GCC region. The scenarios will be based on the current GCC countries *Renewable Energy Landscape*, which depends on the resources, investment opportunities and technology R&D. The main polices to encourage renewable energy projects in the region will highlighted, considering both the unique economic and social characteristic for the GCC. In addition, the scenarios are taking into account the role of the governments and private sectors on driving the renewable energy market in the future. The implementation of either of these scenarios will depend solely on the individual market structure and strength in the GCC countries.

Energy Sources in GCC Countries

The electric power production in GCC countries are heavily dependent on the conventional fuel based technologies, which are very reliable and cheap compared to other alternative technologies. The main source of fuel used in the conventional power stations in the region is natural gas which is produced locally and in some countries is entirely consumed locally for energy production.

However, Many GCC countries natural gas supplies are under the threat of depletion and the countries, except Qatar, may become a net natural gas importer in the near future. The governments are currently evaluating other alternative sources for energy, such as, nuclear power, renewable energy and coal. however, nuclear power is not considered yet a very attractive option due to its both political issues and the possible operational risks or risks related to handling and disposal of Nuclear waste, even if this option was implemented in some of GCC countries, it will be difficult for others owing to technical and environmental consideration. On the other hand, Coal is widely used all over the world, yet the region is poor with Coal mines and it has to be imported, also Coal has a great impact on the environment compared to other current sources of energy in the region. Moreover the introduction of Coal as the main fuel for power production will require the current power stations in the region, which are mainly gas turbines, to be replaced with steam turbine stations.

Electric Power Consumption in GCC Countries

According to the World Bank, the average power consumption for the GCC region reached approximately 11,000 kWh/capita in 2011, which is three times higher than the world average electric power consumption. Kuwait is having the highest consumption per capita then Qatar, whereas Oman recorded the lowest consumption for the same year, as shown in Figure 1.

The GCC region has a very similar annual demand curve which reflects the highly seasonal climate. The average summer demand is more than double of the average winter demand, owing to the increase in residential demand, which is the largest consumer category, during the hot weather in summer. The demand peaks typically in July and August reflecting the highest temperatures and intensive use of air-conditioning.

On the other hand, the water shortage in the region contributes heavily in the energy stress. Yet the GCC region is suffering from water scarcity (less than 1000 m^3/year/capita as defined by WHO[3]), the person share of water consumption is considered the highest in the world figure 3. To ensure clean and adequate supply of water, the governments have constructed, for more than 40 years, many desalination stations, which either operated as a stand alone stations, or as part of cogeneration stations. Two technologies are widely used in the GCC region, based on thermal and Reverse Osmosis (RO) process.

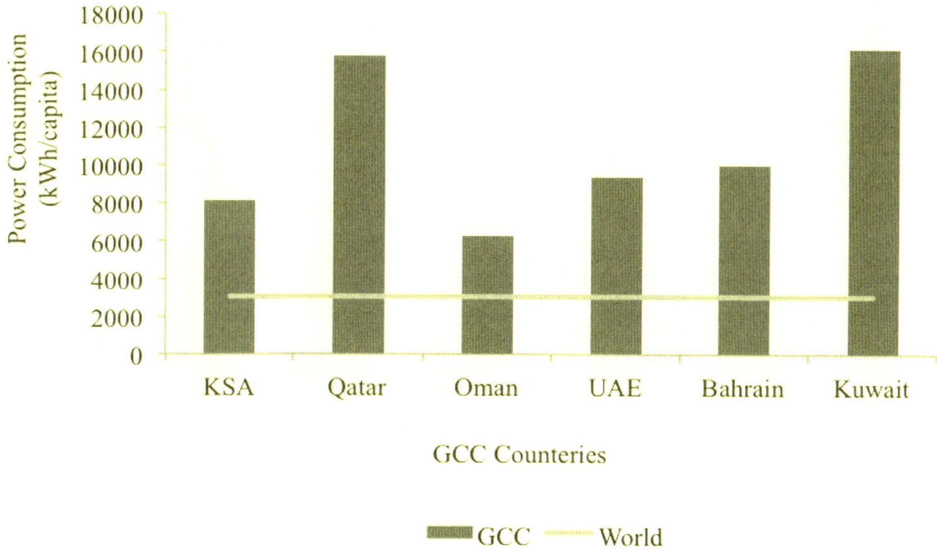

Figure 1 The Average Power Consumption per capita for the GCC and the World (2011)

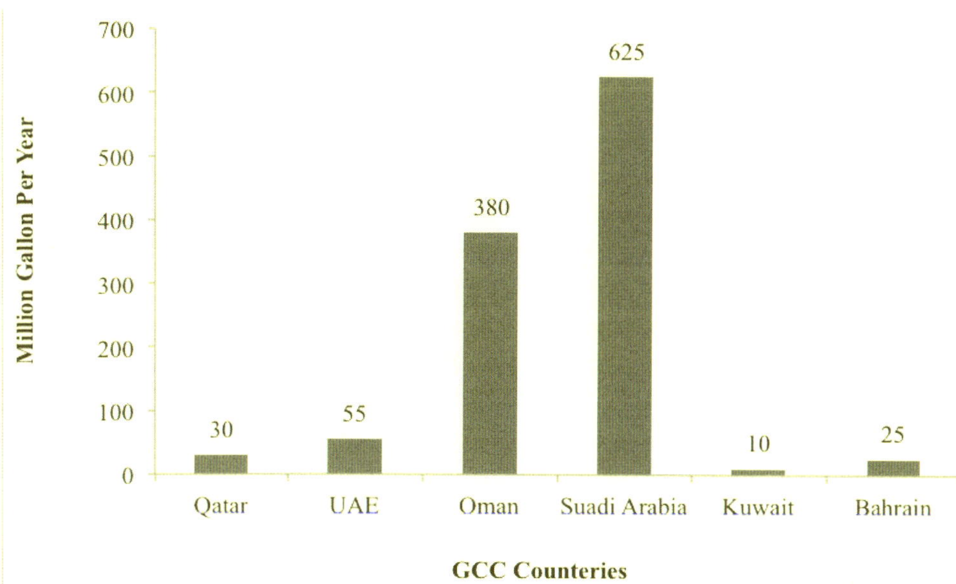

Figure 2 Renewable Fresh Water Resources for GCC

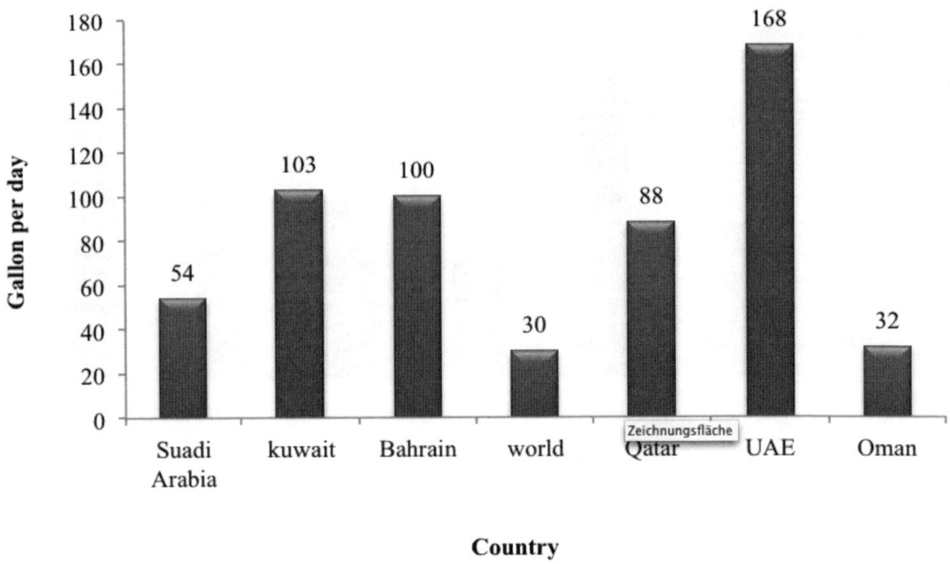

Figure 3 **Water Share per capita in GCC**

Indeed, an enormous investment was conducted for building desalination stations which in some countries represent 90 – 100% of its water resources, more than 50% of the desalination capacity worldwide in operation in 2003 is taking place in the Middle East region, GCC countries and Libya.

Furthermore, Desalination is energy-intensive, so more electricity is needed, causing more pollution from electric power plants. The theoretical minimum power[4] requirement for producing 1 m^3 of sea water is 0.70 kWh/m^3 (Spiegler 1977). In practice, much more power is used (for pumping salt water from the intake, heat losses and air leakage). Even if power were available free of charge, a relatively large investment would be necessary to build desalination plants. When only small amounts of water are needed (small or local systems), these considerations are not of main importance. But in GCC countries the investments are very large and the desalination stations are providing enough water to satisfy the need of their large community.

In GCC the two main desalination technologies that is widely used in large scale water plants is the thermal driven Multi-stage-Flush technology (MSF) and the electric driven Reverse Osmosis technology (RO), figure 4 shows the percentage breakdown of the operating cost for each technology.

The demand projections for both power and water are expected to increase in the near future due to the increase in population and the visions of the GCC countries to encourage different income sources than being only a Oil exporter toward industrialization and tourism, which will require and improve of infrastructure to accommodate the new local and foreign investments in the region, and hence an extensive energy projects will be necessary.

	MSF	RO
Parts	0,016666667	0,065217391
Chemicals	0,083333333	0,152173913
Labour	0,133333333	0,217391304
Thermal energy	0,45	0
Membrane	0	0,065217391
Electrical energy	0,316666667	0,5

Figure 4 **The Percentage Operating Cost for MSF and RO Desalination Technologies**

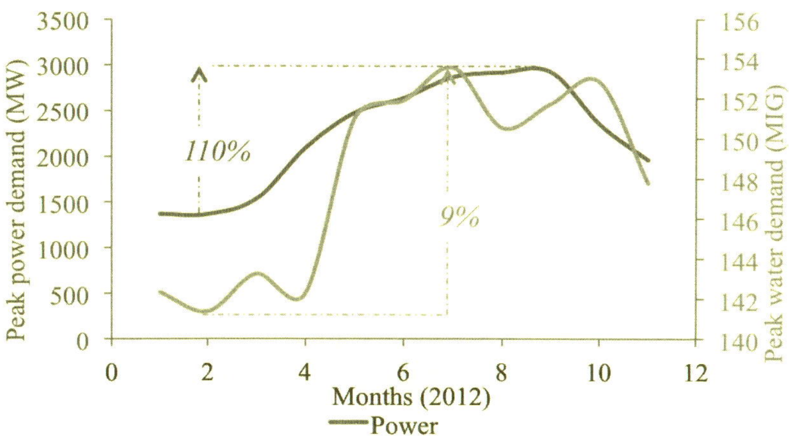

Figure 5 **The Monthly Water and Power Variation (Bahrain)**

The Energy Market in GCC Countries

In the past the power generation, transmission and distribution was owned and operated by governmental bodies (Electricity and Water Authorities) who ensure a reliable and adequate supply of energy. Recently the governments implemented a privatization scheme whereby the electricity and related water assets, liabilities and staff are transferred to private companies. However, not all countries implement the same model or scheme. Some countries implemented the *Integrated Purchasing Agency Model*, where the utility will own some of the generators and the rest will be owned by independent power producers (IPP). Others implemented the *disaggregated Purchasing Agency Model* where all generators are owned by the IPP which will sell the electricity to the purchasing agency which in turn sell to the distribution companies. in the later model the governmental bodies (Electricity and Water Authorities) will secure the provision of electricity and ensure economic operation of electricity sector.

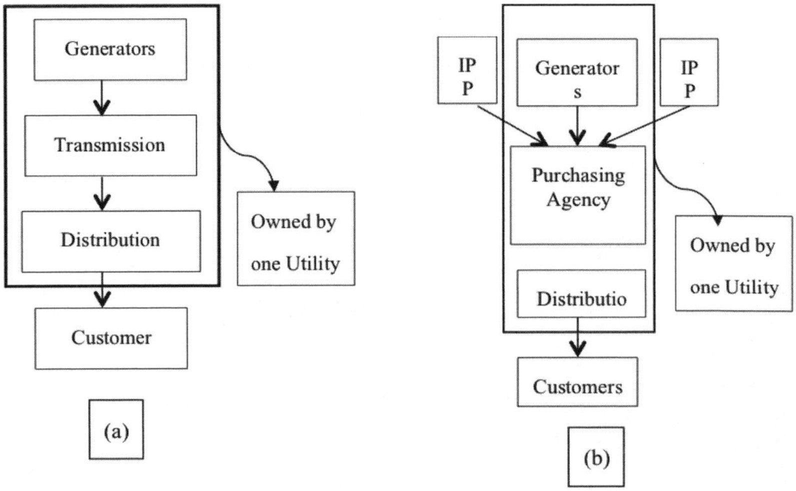

Figure 6 **Schematic showing the Energy Market used in the GCC Countries, (a) Monopoly System Model, (b) is the Integrated Purchasing Agency**

Renewable Energy Resources

Despite the recent interest by the GCC countries in Renewable Energy, the contribution of this source of energy in supplying the electricity demand in the GCC countries is very limited. It is very important to put together political decisions to encourage utilizing the rich available renewable energy resources in the region, especially, that the region has enough financial resources to start large scale projects.

Solar Energy

The first resource of renewable energy that all the GCC countries are rich with, is solar energy. The average annual direct normal irradiation in the GCC region is estimated to reach up to 2400 kW/m^2 (Solargis, 2014). The abundant energy can be utilized in form of thermal or electric energy, using both thermal and Photovoltaic technologies, respectively. However, The lack of accurate long term ground data for solar radiation constitutes a barrier to the construction of a reliable and large scale projects in solar energy. Also the absent of mature R&D in this area, leaves the future of solar energy in the region struggling with technical, environmental and social factors.

Biomass

The GCC countries are considered among the highest countries producing waste in the world, where Bahrain, comes in top of the list, as shown in Figure 7. Biomass resources available in the GCC region can be divided into municipal solid waste and sewage waste.

The solid waste which is mostly household waste end up in landfills, which is large hole in the ground used to bury the non hazardous solid waste under a controlled procedure. Although, landfills are widely used worldwide as safe way to dispose the continuous accumulated waste, it is considered one of the main environmental concerns in this region, especially in countries with limited lands.

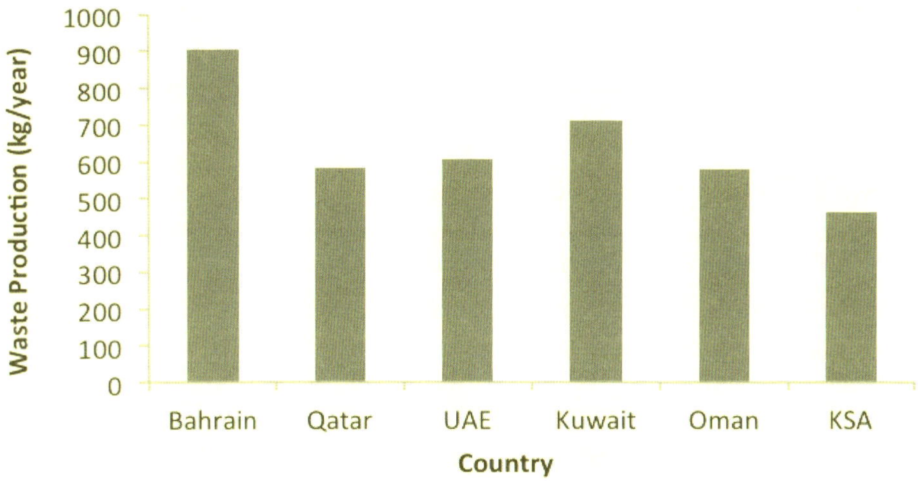

Figure 7 **Annual Waste Production per capita in the GCC Countries**

Landfills can be a very hazardous site, if Methane production and leakage was not controlled, accidents may occur as the temperature of these site increases in summer. Hence, the utilization of Methane gas which is a by-product of the waste decomposition, under heat and pressure, can be utilised to operate a local standalone power station. Also it can be connected to the distribution network and used for residential sector, or small industry. The Utilization of Methane, not only will contribute in electric energy production, but also reducing the Methane hazardous leakage from these sites.

Utilizing the produced gas will require a comprehensive surveying to the new landfill site, and a research group to analyse the decomposition process under the weather and ground characteristic of the GCC region.

On the other hand, Methane can also be harnessed from the sewage water treatment plants and contribute to the total Methane production from Biomass. In fact, in the GCC region, producing the gas from the sewage water treatment is more controllable and predictable than Landfill.

Wind Energy

Although the GCC region is not part of the Global wind zone, it has an average wind power of 650 W/m^2. Despite solar energy, wind in the region is highly dependent on the site characteristic, and hence different areas in the region will have different wind energy potentials, Figure 8.

The region can utilize wind power for both small and large scale projects using the right wind turbine types which will be suitable for the wind speed characteristic in the region.

Similar to solar energy, wind energy will not be able to fully be investigated unless, a long term wind data at different hub height is collected for different potential sites in the region. Also, integrating wind energy into the current power system will require a comprehensive analysis to ensure system stability due to the variable nature of wind energy.

Renewable Energy Projects Opportunity

Renewable Energy Desalination Systems

Desalination technology is well developed and proven, but does consume a considerable proportion of the country energy resources. As reserves of fossil fuel become depleted and global concerns regarding CO_2 emissions increases, it is considered attractive to harness the renewable energy, especially solar energy, to operate the desalination units.

In 1984 a one million Dollar thermal desalination station coupled with solar thermal collectors was constructed in Um Nar, UAE, it was the world first largest renewable energy station with a capacity of 120 m^3/day. The station was decommissioned in 2003 as a consequence of privatization of the water and power sector in UAE (El-Nashar and

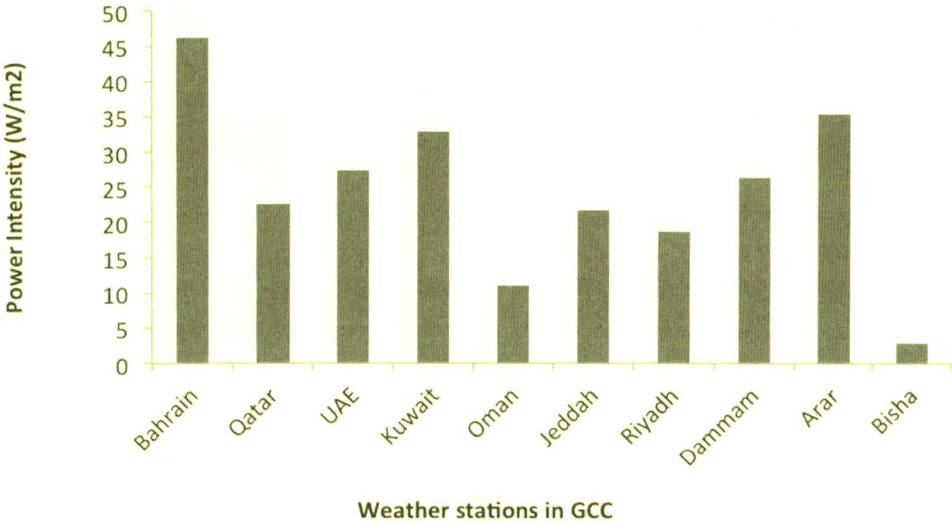

Figure 8 Average Annual Wind Power Intensity

Al-Baghdadi 1998; El-Nashar and Samad 1998; El-Nashar 2003). Most of the station downtime was due to problems with evaporator pumps and their motors.

Actually, the region had number of small systems installed before Um Nar station, where they could not survive more than two years; the most famous station is Yanbu solar freezing desalination which was installed with capital cost of 35 million dollars in 1984 with a capacity of 200 m^3/day and decommissioned in 1986 for financial problems.

The study to combine other renewable energy technologies with the desalination systems was the main objective of many researchers, which leads to a number of small to medium sized systems, designed, modeled and installed all over the world, especially the Middle East, the Mediterranean, Australia and different remote area and laboratories. However, the current systems in operation are not what anyone will expect. Many stations were decommissioned, and others are still inside a designing model or stored in a small lab.

The interest to combine other renewable energy technologies with the desalination systems was the main objective of many researchers, which leads to a number of small to medium sized systems, designed, modeled and installed all over the world, especially the Middle East, the Mediterranean, Australia and different remote area and laboratories. However, the current systems in operation are not what anyone will expect. Many stations were decommissioned, and others are still inside a designing model or stored in a small lab.

The papers will be categorized according to the type of renewable energy technologies. As shown in figure 9, considerable amount of research were conducted in solar energy (86%), focusing in particular in solar still, where only 14% are the share of all the other technologies.

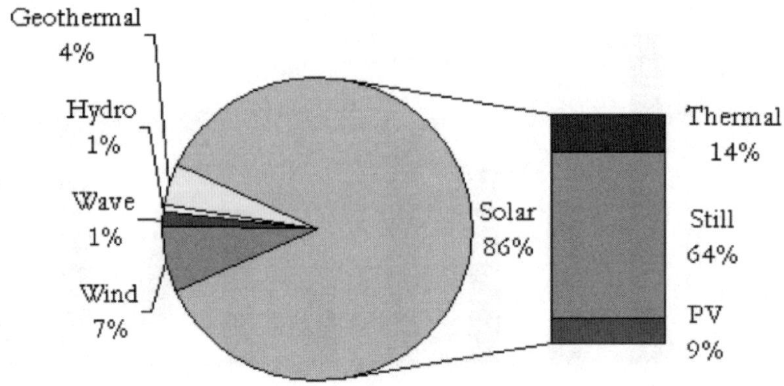

Figure 9 **The Percentage of the Published Scientific Papers in Renewable Energy Desalination**

Comparing the total number of the published papers, 47% was done in Middle East and only 1% are currently running, most of which are small RO systems for brackish water, Figure 10. The research on coupling renewable energy and desalination conducted in Middle East, for last 50 years, had a minimal contribution to encourage the establishment of a serious projects in the GCC region in specific and Middle East in general.

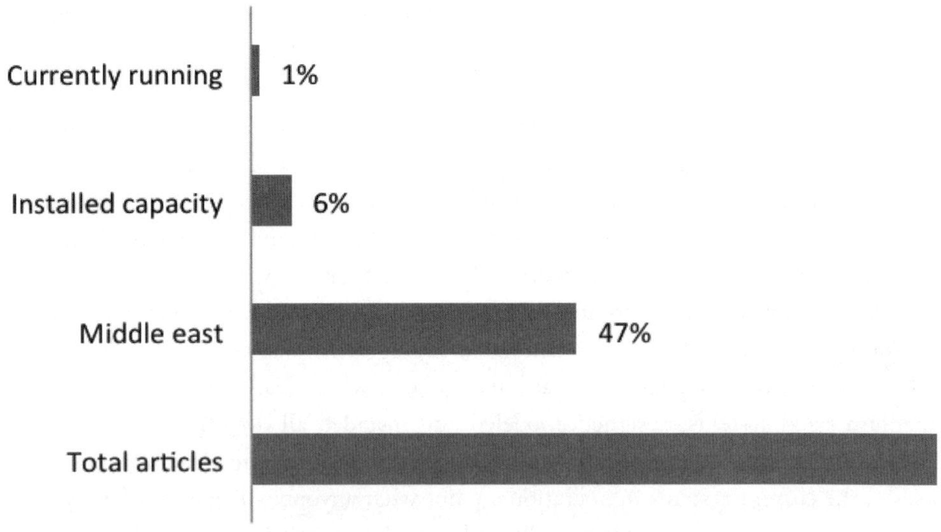

Figure 10 **Currently installed Renewable Energy Systems in the Middle East**

The main reasons for the non continuity of operating any installed renewable energy powered desalination systems in GCC countries, can be summarized in five points:

1. The organization or the project group not anymore in charge of the project.
2. Privatization
3. Maintenance
4. Finical ....!
5. Not any more interested in the project ....!

Solar Energy for Electric Power Production

One of the main challenges of using solar panels in the GCC regions, is the accumulation of dust and the need to maintain their performance through a continuous cleaning, and consequently will increase the solar energy cost in the region.

On the other hand, the cities structure in the GCC region is very attractive for the application of solar panels roof installations. The cities are characterised with flat roof buildings which is usually one or two stories in the residential area.

Using residential buildings roof tops will provide the following:

- required area for solar panels installation.
- sand storm are less frequent compared to the open sites, hence, better solar panel performance and less maintenance cost.
- ensure no shading on the installed solar panels by other nearby buildings or trees.
- access to power grid, through connecting solar panels to the distribution system.

Renewable Energy Policies

The Gulf region can be the future market for a large investments in renewable energy, which will open a great opportunities for renewable energy technology developers to improve the current technologies to suit the region climate.

In this chapter two scenarios have been proposed keeping in consideration the economic and social sensitive characteristic of the region. The two scenarios are sharing four main policies, as follows:

1- Technical Policies

- Identifying and setting both distribution and transmission network infrastructure requirements to accommodate large renewable energy integration in the system.
- Setting conventional power stations an operation policy to minimize the impact of renewable energy on the station output.
- Introducing GCC standards for renewable energy systems installation.

2- Energy Efficiency

- Regulations to enforce the law of using energy efficient devices.
- Implementing different tariff categories.
- Monitoring electricity usage for buildings and maintaining regular inspections to buildings with high electricity usage.

3- Awareness

- Include the energy conservation in the school curriculums.
- promote clean energy technologies among individuals and the industry sector.
- Establish university degrees or compulsory courses in energy efficiency, conservation and renewable energy in the GCC universities.

4- Research and Development

- Investigate the region renewable energy resources through the installation and recording a long term data.
- Funding the research in energy related issues in general and renewable energy in specific.
- Thoroughly studying the impact of integrating renewable energy into the power system and investigate impact on the GCC interconnection.
- Establish collaboration among the GCC countries through introducing a regional renewable energy research centre for the GCC countries which include universities from the region as members.

Some of these policies have been implemented in some of the GCC countries, and yet it need to be enforced in others.

Scenario I: The Government-Oriented Scenario

In this scenario the government will have the full responsibility to introduce the renewable energy into the energy market at the region. The scenario will include both monitoring the market and providing the fund to the renewable energy projects.

Although the energy sector in the GCC countries is privatized, it is still largely government oriented market, where the private sector plays modest role, which is limited to operating the power stations which is owned by the governments.

In this scenario, the renewable energy projects which will be totally financed by the government and operated by private companies will be constructed on a government owned lands or rented lands from individuals.

In case of the solar panels on roof tops project, for example, three policies will be introduced:

1- Rental pricing and duration.
2- Installation requirements, which will include the suitability of the building for solar panel installation.
3- Maintenance duration and accessibility. Number of maintenance per month and duration should clearly be specified depending on the characteristic of the site and the season.

Scenario II: The Market-Oriented Scenario

In this scenario, the private sector will have a larger role than the first scenario. The private sector, will have the full responsibility to carry a resource assessment for potential renewable energy projects at any site. They also will conduct a full feasibility study for the RE projects, contact land owners and technology providers

The Electricity utility, which represent the government in this case will ensure the connection of the renewable energy system with the national grid, and maintain the overall system stability.

In case of the solar panels on roof tops, the private sector can contact the roof owners directly for installing solar panels, the companies can be local or foreign which is regulated through number of policies as follows:

1- Installation licence.

   The companies involved in installations must have a licence issued from a local governmental body.
2- Power purchasing contract.

   The companies must have a purchasing agreement with the power utility before the installation.
3- Incentive to encourage the private sector to invest in renewable energy.

Conclusion

The future of the renewable energy in the GCC countries depends on the approach that these countries will consider in order to promote the renewable energy and minimize the dependence on the fossil fuel. The high energy subsides in the region and the lack of robust policies, will limit the role of the end user (consumer) to energy conservation and devices efficiency.

The two main players in the renewable energy marker in this energy is the governments and the private sector. This chapter has presented two scenarios, considering the two major players, and the implementing of either of the two scenarios, will depend on the strength and structure of the current market on each GCC countries. These scenarios should encourage the research in the region to be directed toward the renewable energy market in order to come out with solutions to the current technical obstacles that will affect the performance of the renewable energy system and their integration to the available power system.

Notes

1 World Energy outlook 2013 fact sheet.
2 The Renewable Energy Country Attractiveness Index, Ernst and Young, 2012.
3 World Health Organization.
4 Is any sort of energy that can be completely converted to mechanical work to desalinate a ton of sea water (34,300 ppm).

References

Al-Amir, J. and B. Abu-Hijleh (2013). "Strategies and policies from promoting the use of renewable energy resource in the UAE." *Renewable and Sustainable Energy Reviews* **26**(0): 660-667.

Al-Badi, A., et al. (2009). "Power sector of Oman—Today and tomorrow." Renewable and Sustainable *Energy Reviews* **13**(8): 2192-2196.

El-Nashar, A. M. (2003). «Effect of dust deposition on the performance of a solar desalination plant operating in an arid desert area.» Solar Energy **75**: 421-431.

El-Nashar, A. M. and A. A. Al-Baghdadi (1998). "Exergy losses in a multiple-effect stack seawater desalination plant." *Desalination* **116**: 11-24.

El-Nashar, A. M. and M. Samad (1998)."The solar desalination Plant in Abu Dhabi: 13 years of performance and operation." *Renewable energy* **14**(1-4): 263-274.

Malik, A. S. and S. Al-Zubeidi (2006). "Electricity tariffs based on long-run marginal costs for central grid system of Oman." *Energy* **31**(12): 1703-1714.

Mezher, T., et al. (2012). "Renewable energy policy options for Abu Dhabi: Drivers and barriers." *Energy Policy* **42**(0): 315-328.

Reiche, D. (2010). "Renewable Energy Policies in the Gulf countries: A case study of the carbon-neutral "Masdar City" in Abu Dhabi." *Energy Policy* **38**(1): 378-382.

Solar, G. (2010-2014, 2014). "Solargis." Retrieved 17/6/2014, 2014, from http://solargis.info/.

Spiegler, K. S. (1966). *Principle of Desalination*. New York, Academic press.

9

The Geopolitical Shift of Energy Supply and Demand: The Implications on the Middle Eastern Energy Suppliers

Sara Vakhshouri

Introduction

The shale revolution in the U.S. is changing the country's position in the market from a major importer of energy to an energy exporter. In 2013, U.S. energy production reached to 81.7 quadrillion BTU[1] (quads) which was sufficient to supply 84% of its total energy demand of 97.5 quads.[2] In the same year, U.S. energy mix included 82% of fossil fuels, 10% renewable and 8% nuclear energy. U.S. domestic energy supplies, particularly natural gas and crude oil have, increased markedly since 2005.[3] Between 2005 and 2014, U.S. tight oil and shale gas production boosted significantly and rose nearly 34 and 65 percent, respectively.[4]

American oil output in 2013 has witnessed its largest annual increase since 1989, reaching 7.44 million barrels per day. It also projected that by 2015, America's oil output will reach to 9.6 million barrels a day – the highest number ever produced since 1970.

Dry natural gas production in 2013 was also at its highest level ever, having increased by 21 percent since January 2010. Most of this production increase has been a result of increase in drilling productivity particularly in Marcellus shale formation. Alongside this production increase, U.S. exports of refined petroleum products and natural gas pipeline export to its neighboring countries grew dramatically last year. Refined petroleum exports jumped 10 percent in 2013 and reached to 2.95 million barrels a day in October of this year.[5] This was a 70 of percent increase in comparison to the past decade. U.S. natural gas pipeline exports to its neighboring countries in 2013 peaked at nearly 99 billion cubic feet (bcf) in January. This number was 84 bcf in 2012 and 25 bcf in 2003. Over the past decade, U.S. natural gas exports to Mexico, via pipeline, almost doubled and more than tripled to Canada. At the same time, overall U.S. energy imports in 2013 fell by 10 percent compared to 2012, and 15 percent in compared to 2011.

The last time this ratio of U.S. domestic energy supplies to consumption was so high was during the 1978 to 1982 period. This was the time that U.S. oil consumption declined as a result of global oil prices rose due to the 1979 revolution in Iran, and the Iran-Iraq war between 1980 and 1988.

This dramatic increase of domestic energy production is, as we know, mainly the result of technological innovations. Hydraulic fracturing and horizontal drilling technologies have increased U.S. oil production by 30 percent. At the same time, U.S. domestic demand for petroleum and coal has fallen due to increases in vehicle efficiency, declining road travels, and the usage of other fuels in power generations. This increase of domestic production versus consumption has, crucially, reduced U.S. dependency on energy imports from its traditional energy suppliers, mainly the Africa. In May 2013, U.S. domestic crude output exceeded its imports for the first time in 16 years. U.S. domestic oil inventory also jumped to its highest level in 82 years, while its imports from OPEC member countries plummeted by 26 percent.

In theory, U.S. energy production, and its possible supplies, could greatly influence global energy market and prices. In effect, the U.S. stands poised to become a major shaper of global energy security, this time from the position of an exporter. This chapter, therefore, discusses the effects of the shale revolution in the U.S. on the global energy market and its possible implications on the Middle Eastern energy suppliers over the next decade.

Global Energy Outlook

In the next 20 years, population and wealth will remain the major factors in determining world energy demand. Over the last 20 years, the world's population increased by 1.6 billion.[6] It is projected that by 2040 the global population will grow to nearly 9 billion, up from 7 billion in 2010, and overall global economic activity will double before 2030. [7] Growth in world GDP is expected to increase to 3.7 percent per year over the next 20 years, up from 3.2 percent per year between 1990 and 2010.[8] During the same period, global energy efficiency is also expected to grow at rate of 2 percent per year, almost double than the growth rate in the past 10 years.[9]

The acceleration of higher energy efficiency, combined with higher energy prices, will put pressure on overall energy demand growth. It is projected that energy consumption growth will have a bearish trend and decelerated to average 1.6 percent per year between 2010 and 2030, down from 2.5 percent per year over the past decade. This rate will decline to 2 percent per year until 2020 and will move downward even further to 1.3 percent per year between 2020 and 2030. The total consumption growth of average 1.6 percent per year between 2010 and 2030, will add 39 percent to the global energy demand by 2030.

Non-OECD economies will drive the major share in the growth of energy consumption by 2030, with an average growth rate of 2.7 percent annually. These countries will constitute 96 percent of energy consumption growth rate by 2030, up from 69 percent in 2010. Between 2010 and 2030, non-OECD consumption will account for 65 percent of global energy demand, up from 54 percent in 2010. China, India and the Middle East will account for the major non-OECD consumption growth, with 8, 3.5 and 4 million barrels per day of increase by 2030, respectively. Conversely, the energy consumption of OECD countries will grow at a tepid rate of 0.2 percent annually until 2030.

While oil, gas and coal preserve their major share in the fuel mix, renewables, including biofuels, will have the fastest growth rate of 8.2 percent per year. Oil will have the slowest growth of 0.7 percent per year by 2030. During the same period gas will have the largest growth among fossil fuels, around 2.1 percent per year, and will slowly substitute oil and coal.[10]

On the supply side, though, we see a much different story. Global energy supplies are expected to grow at a faster pace, particularly due to the accessibility of unconventional resources. Here, we see major growth coming from conventional crude oil and natural gas liquids (NGL) of OPEC countries, particularly from Saudi Arabia and Iraq. By 2030 there will 12 million barrels per day of additional supplies from OPEC countries and 5 million barrels per day of additional supplies from non-OPEC producing countries. The major production raise in non-OPEC countries will be from U.S. tight oil, Canadian oil sands, Brazilian deep-water resources, and U.S. and Brazilian biofuels.

Shale Gas and Tight Oil Production

Access to unconventional resources that were once uneconomic is now possible with the help of new technologies. These new technically recoverable resources could change the overall outlook and functioning of global energy supplies. According to the EIA, 32 percent of recoverable global natural gas and 10 percent of global recoverable crude oil are located in 137 shale formations in the United States and 41 other countries.[11] According to this assessment, there are 10 other areas in addition to U.S. that have large shale gas and oil resources. The top five countries with recoverable tight oil resources are Russia, U.S., China, Argentina and Libya,[12] while China, Argentina, Algeria, the U.S. and Canada are the top five countries with recoverable shale gas formations.[13] (See Table 1 and 2)

However there is no guarantee that technically recoverable resources would be also economically recoverable. For resources to be recoverable economically, there needs to be a profit incentive in their production under current market conditions. Different factors of volume of produced natural gas or oil from a well over its lifetime, the cost of drilling and maintaining a well, combined with global prices, all determine the bottom-line profitability of recoverable resources. Current evidence suggests that the profitability of natural gas and crude oil production from shale formations in the U.S. might not apply in other countries. Particularly in the case of shale gas and tight oil, the complexity of required technology and infrastructure reduces its profitability in countries outside the U.S.

Currently, the U.S. and Canada are the only major producers of tight oil in the world. The U.S. accounts for 91 percent of tight oil production from North America (See Chart 1).

In 2013, the U.S. produced an average of 2.5 million barrels a day of tight oil and Canada produced about 340 thousand barrels a day. The share of tight oil was almost about 10 percent of Canada's total oil production of 3.52 million barrels a day. Specifically, major tight oil production in Canada comes from Alberta, Saskatchewan and Manitoba. If we move to Eurasia, we see similar patterns. Russia, with its vast West Siberia Basin, produces little in tight oil. In 2013, it produced a mere 0.12 million barrels of tight oil per day – only 1 percent of its total oil production.

As mentioned earlier, unconventional resources are a different story in the U.S. In America, technically recoverable shale resources are produced in larger volumes and, crucially, at relatively lower costs. The shale gas and tight oil resources have been significantly altered this country's energy outlook. Until the early 1970s, the U.S. was the largest oil producer in the world at 9.6 million barrels a day. However as it consumption was growing, its production declined until 2008 when it dropped to 5 million barrels a day. Yet since 2008, the usage of new technologies of horizontal drilling and hydraulic fracturing allowed the U.S. to produce oil and gas from its shale formations. In the last quarter of 2013, U.S. tight oil production reached 3.22 million barrels a day and raised its total oil production to 7.8 million barrels a day. This drove U.S. total crude oil production to about 10 percent of global world production, up from 9 percent in 2012. (See Chart 2)

Most crude oil production increase is United States is from the two shale basins of Bakken in North Dakota and Montana and Eagle Ford in Texas. In February 2014 Eagle Ford produced about 1.21 million barrels a day or 36 percent of U.S. total oil production. In the same month, Bakken's share was 28 percent of America's total production, produced 0.94 million barrels a day. Because of these crucial resources and areas, overall U.S. production is expected to rise to around 5 million barrels a day by 2017 and reach its historical peak of 9.6 million barrels a day by 2019.[14] It is also bullishly projected that this trend will put the U.S. above Saudi Arabia as the largest crude oil producer in the world with production of 11.6 million barrels a day.[15] Likewise, shale gas production is also expected to grow sharply and U.S. natural gas will be exported by pipeline and also as liquefied natural gas (LNG). By 2035 natural gas will surpass coal and will be the largest source of American electric power generations.[16]

Tight Oil Production and Price Sensitivities

The extraction of unconventional resources not only requires high technological proficiency, but it is also very costly in compared to conventional production. The main reason for this is due to the low efficiency of shale oil wells. The production from an unconventional well reaches to it production peak at the early stages of operation and its production drops dramatically after first few weeks of drilling. Therefore in order to maintaining the tight oil production level, unlike in the conventional well, there is a need for regular drilling of new oil wells. (See table 3). Bluntly put, no country expect other than the U.S. has undertaken such massive, and sustained drilling operations for the extraction of oil and gas from unconventional basins.

Most of America's unconventional oil resources are profitable at the WTI[17] price range of $50 to 65 per barrel. In a short-term scenario, if global oil prices remain above $65 per barrel, U.S. would be still able to increase its tight oil production to 5 million barrels a day by 2017. Thus the production from shale basins will drop sharply if the global oil prices fall below $65.[18] For most of the U.S. tight oil resources to be economically developed and produced, oil prices should remain at least around $70 per barrels in long-term. The production from shale and tight basins is highly sensitive and resilient to the prices fluctuations. The minimum oil price for maintaining the tight oil production from Bakken basin is $60 per barrel and is $80 per barrel in the Eagle Ford shale formation.[19] With current costs, it is expected that overall tight oil production will drop about 20 percent with a downturn of oil prices below $70 per barrels. The main reason for this considerable production drop rate is that under this price range, no new well would be drilled to maintain the production level. If oil prices drop bellow the economically profitable production, the drilling of the new wells, for maintaining production levels, will stop and the tight oil production reduces significantly with in a period of between 3 to 6 months. Thus, perhaps ironically, the reduction in tight oil supplies will have an upward effect on the prices. Therefore the range of $70 to $80 per barrels is considering the minimum price range that market could sustain its balance of production. Thus, the relationship of tight oil production to oil prices and also the reaction of other major oil producing countries, particularly OPEC members, to the any steep price dip, will have a significant role in offsetting the oversupplied oil in the market by maintaining the prices at least at the minimum range.

Implications of Unconventional Resources on the Middle East

U.S. unconventional energy production and exports could have different effects on Middle Eastern producing countries. The volume of American oil and gas production, exports, and also prices will foster intense competition for the energy market's 'traditional players'. One of the core questions in this paper is how U.S. unconventional energy production and exports could affect the Middle East? These effects are defined in different categories of oil prices, imports from the Middle East, and general competition in the energy market.

Oil Price Impacts on Middle East

As mentioned earlier, new technologies and accessibility of unconventional energy resources have rewritten most of supply predictions made during the past decade. Contrary to most of the global supply projections of a decade ago and theories like peak oil, the global energy production is increasing with higher rate than the global demand growth. The surge in world energy production could surpass the global consumption and have a steep impact on the prices.

Tight oil production has not yet had a downward effect on global oil prices, mainly due to supply interruption in major oil producing countries such as Libya and Iran. In 2012, tight oil production added 847 thousand barrels per day to U.S. crude oil output. However this did not lead to a significant price drop. Over the past few years, the 'tight oil revolution' is solely a U.S. phenomenon and has not proved easily replicable in other countries. U.S. production in the mid to short-term will not be substantial enough to impact global oil prices significantly. Indeed, it may take a long time before global oil prices are shaped by tight oil production. Nonetheless in the long-term the story could be different and the market could, quite possibly, face the prospect of oversupply.

However there are other factors combined with North America's tight oil production that could soften the global oil prices. One major factor is lower demand growth due to higher energy efficiency, high oil prices and lower economic growth in some regions such as the EU. The other factor is higher oil production capacity. Additional oil production capacity between 2010 and 2020 is expected to be 17.6 million barrels a day – the highest increase in any decade since 1980s.[20] Production capacity is expected to grow even further in then next decade because of unconventional oil production in U.S., Canada and Brazil, combined with conventional oil production in Iraq. Iraq is currently producing around 3.5 million barrels a day of oil [21] and it is expected that this will rise to 4-4.2 million barrels a day in the near future. Libya also could regain its disrupted oil production to about 1.3 to 1.4 million barrels per day. Iranian crude oil production, hit hard by international sanctions, could potentially rise by 1 million barrels a day. Therefore, the combination of the growth in oil production capacity, and lower demand growth, could potentially plunge oil prices by the end of the current decade.

Any downward price movement in the market would undoubtedly have a significant effect on Middle Eastern energy producers, their economies, fiscal regimes and political stability. As we know, this is a result of their fiscal dependency upon oil revenue. For instance, Saudi Arabia, a country with largest conventional oil reserves of almost 266 billion barrels[22] and the largest producer of petroleum liquids is strongly dependent on its oil revenue - in 2011 petroleum exports generated 90 percent of its overall revenue.[23] Kuwait is another example. With crude oil production of 2.9 million barrels per day, and exports of 2 million,[24] petroleum exports accounted for 93 percent of Kuwait's export revenues and 95 percent of its fiscal income.[25] Therefore the expected dip in the prices means less revenue for Middle Easter oil producers and could have negative impact on their economies, social spending and, possibly, political stability. Yet lower energy prices could, ironically, possibly benefits these countries. Domestic energy prices in these countries are highly subsidized. With a decline in energy prices, these governments would have a shorter gap to cover as the energy subsidies in the both for power sector and transportation.[26]

However, as mentioned earlier, its not expected that global oil prices will drop below the \$70-\$80 per barrel range. This is mainly because of the vulnerability of tight oil production to global market prices. Indeed, if tight oil supplies in the market plunges significantly due to lower oil prices, the market will react to supply interruption and prices will rise again. With OPEC members highly dependent on oil revenue, they could also balance their production in order to offset potential oversupplies in the market and, in effect, ward off steep drops in prices.

Crude Oil Imports from the Middle East

U.S. imports of crude oil and natural gas have shrunk due its gas and oil production from shale formations. The tight liquid production increase from the Bakken and Eagle Ford shale basins in North Dakota and Texas over the past five years was, in fact, equal to Iraq's current production. The Bakken shale formation itself could be one of the major Persian Gulf oil producing countries inside of U.S.[27] Thus, the growth of U.S. shale gas and tight oil production is expected to continue and U.S. dependency on energy imports will fall significantly over the coming decades.

To understand the impact of American tight oil production upon their imports from Middle East, we must first carefully consider the specifications of the crude oil used as feedstock in domestic refineries. Gravity or API degree (heavy or light crude oil) and sulphur content (sour or sweet) are the two major elements determining crude oil's type.

The gravity or density of the crude oil is based upon the American Petroleum Institute's (API) definition. The API degree specifies the density of the crude oil; higher API degree indicates the lower density (higher number of light hydrocarbons) or light crude oil. According this scale, crude oil with API greater than 31.1 degree is known as light. The range of 22.3 to 31.1 degree is considered medium, and less than 22.3 degree is heavy crude oil. An API degree below 10 is categorized as extra heavy crude oil, while oil with sulphur content of above 1 is considered sour.

American refineries are mostly refining heavy (in the range of 20 to 21 degrees of API gravity or lower) and sour (sulphur content of above 1 percent) crude oil as their feedstock. Saudi Arabia, Venezuela and Canada are major suppliers of heavy crude oil. Nevertheless, U.S. domestic tight crude oil is mostly light (with API gravity of above 35 degree) and sweet (with sulphur content of less thane 0.5 percent).[28] Therefore increases in crude oil production from the tight formations result in a reduction of light and sweet crude oil imports. At the current time, 50 percent of U.S. refinery inputs is supplied by domestic crude oil production. In the first quarter of 2014, U.S. average imports of light and sweet crude oil from Africa was about 170 thousand barrels per day, down from 375 thousand barrels per day in the third quarter of 2013, and down from 2 million barrels day in 2010. Most U.S. light crude oil has historically been from Africa, primarily Nigeria, Angola and Algeria. EIA's projections indicate that by 2040, U.S. imports of crude oil will shift toward heavier crude oil with API gravity of 25.6 degrees from 27.6 degrees in 2012.[29]

This not only indicates that the U.S. would still be dependent on heavy crude oil imports from Middle East, but also that imports from this region will rise in the next 2 to 3 years. The major factor for increase in Middle Eastern imports is a larger Venezuelan supply shift to Asian markets. Venezuelan crude oil has traditionally been one of the major heavy crude oil supplies for the U.S. Yet Venezuelan exports have recently redirected towards Asian markets, with new commitments in India and China. Venezuela has recently signed a term contract[30] for supplying 500 thousand barrels per day of crude to India.[31] Venezuela is currently supplying 300 thousand barrels per day of crude oil to China, and this is expected to grow in the near future.[32] Therefore, as a result of Venezuelan shift to Asian market, combined with its limited production capacity, Middle Eastern (particularly from Saudi Arabia and Kuwait) exports to the U.S. have actually increased. This will more likely continue until the policy limitations on Canadian very heavy oil sand (with 20 to 21 degrees of API gravity) are lifted. Under this scenario, Canadian oil sand could gradually substitute Middle Eastern supplies. Nevertheless, anther major paying factor in importing oil from the Middle East is quite simply, price. Canadian and Venezuelan prices are higher than Middle Easter ones. Substituting the 25 percent of Middle Eastern oil will prove more costly for American refineries.

Even under a scenario where Canadian oil sand flows to the U.S. market, Middle Easter supplies will never stop completely, or all that significantly. It should not be forgotten that Saudi Arabia has a refinery capacity of 1 million barrels a day. Moreover, Saudi Aramco has a 50 percent share of Motiva Port Arthur refinery in Texas along with Royal Dutch Shell (with also 50 percent share). This refinery has the largest capacity to produce gasoline and diesel in United States. This allows Saudi Arabia to maintain its minimum market share in the U.S.[33] Here, infrastructural considerations trump market fundamentals.

Competition in the Market

U.S. unconventional productions pose a major competition for Middle Eastern countries in global energy market. U.S. shale gas, refined petroleum products and petrochemical products will, above all, challenge Middle Eastern supplies in the their traditional markets. Let us proceed to examine a few specific cases.

Natural Gas – In 2012, the shale gas production of the state of Pennsylvania alone was equal to total natural gas export of Qatar, the second largest natural gas exported in the world.[34] Yet current evidence suggests that U.S. LNG[35] exports will soon eclipse both Qatar and Australia as the largest exporters in the global market. U.S. gas supplies will increase flexibility in the natural gas and LNG markets, and create severe competition for LNG exporters. American's LNG pricing and volume of exports will greatly affect the market. On the other hand, and perhaps ironically, the Middle East could conceivably become a major market for American LNG. United States could supply some portion of the natural gas demand in Kuwait and UAE,[36] especially given that Iran's natural gas exports are negligible.

Refined Petroleum Product Exports – the U.S. has become a significant competitor to Middle Eastern refined products, particularly in the European market. In 2008, the U.S. exported around 1.6 million barrels of refined petroleum products. This number hiked to above 2.7 million barrels a day in 2013 (See Chart 3).

Petrochemical Exports – The volume of shale gas production creates a great advantage for U.S. gas based petrochemical products in comparison to Middle Eastern products. Most of the gas based petrochemical producers in the Middle Eastern lack enough gas as a feedstock for their petrochemical factories. Saudi Arabia, one of the major gas based petrochemical producer in the Middle East, is using LPG[37] and Naphtha as a petrochemical feedstock in order to be able to offset the shortage of natural gas in its power sector. Thus, the U.S. has a clear, and long-term advantage over the Middle East in the gas based petrochemical industry and exports.

Unconventional Resources and the New Geopolitics

The new shale gas and tight oil resources in the U.S. could precipitate a major paradigm shift for the global oil market. By extension, this could have major effects on global geopolitical situations and relationships. One way to gauge the impact of these new supplies, through a prism of global energy security, would be to examine the rivalries between major oil and gas producers in their bids to secure markets and security of supplies. The alterations in traditional markets and energy trade partners as a result of changes in the flow of energy could itself create a whole new notion of security and strategic partnership between different countries and regions. Indeed, the very standing paradigms and understandings of global energy security may need to be reassessed.

Yet conversely, the extraction of gas and oil from unconventional basins is a very new and recent phenomenan. Markets are still uncertain about their development, production trend and the broader impacts on the global energy market. The U.S. is still in the process of evolving a new strategy regarding its energy resources. The investment outlook, potential markets and question of removing the ban on oil exports are still under debate among experts and policymakers. Clearly, the U.S. view of the potential impacts of its unconventional energy resources on the global geopolitics, coupled with its own national security concerns, will shape its strategy of incorporating such resources.

Whatever the outcome of U.S. energy policy becomes, the shale boom in North America has clearly, and greatly affected the dynamics of global energy stability, energy market, and pricing. It will also change the flows of energy trade and the relationships between traditional suppliers and consumers. Therefore projections of shale liquid production trajectories, and their impact on prices and energy flows, have changed the perspectives of every major player in the energy market. Major energy consumers such as China, India and Japan, and major energy producers such as Saudi Arabia and Russia, have all started rethinking and revising their energy policies and strategies. The emergence of new supplies from unconventional resources could also have great impact on countries like Iraq, Mexico and Iran, which that are struggling to increase their production and regain lost production capacity. The shift of supply and demand, and the flow of energy trade, could change key relations between Middle Easter supplies and their customers. Therefore even on the foreign policy side, these countries have started to reshape and shift their foreign policy priories accordingly. The bottom line is that the existence of the new unconventional resources could change the dynamics of global energy market and influence the traditional market players and, conceivably, create a new geopolitical landscape.

Conclusion

The technological developments and the recent shale revolution in North America have significantly changed the global energy landscape. The U.S., once was seen as major market for oil and gas exporters, is now expected to outstrip Saudi Arabia in oil production and surpass even Qatar and Australia in LNG exports. Energy projections demonstrate that overall U.S. oil production by 2015 will reach its historical high level of 9.6 million barrels per day in 1970. It is also expected that U.S. tight oil production will surge to five million barrels by 2017.

Current evidence strongly suggests that there are 10 other countries that possess large shale gas and oil resources. The top 5 countries with recoverable tight oil resources are Russia, the U.S., China, Argentina and Libya. China, Argentina, Algeria, U.S. and Canada are top five countries with recoverable shale gas formations. However, the high complexity and the costs of shale gas and tight oil production have made the shale boom largely American phenomena, at least in the short term.

Unlike conventional oil wells, there is need for regular drilling of oil wells in the unconventional formations to in order to maintain the production level and offset the production loss from old wells. Tight oil production is extremely sensitive to global oil prices, due to the low efficiency of unconventional oil wells coupled with the high costs of drilling. Oil prices should stay in the range of $70 to $80 per barrels for shale oil production to continue its growth and maintain profitability. If crude oil prices dropped to under $70 per barrel, with current production costs, U.S. tight oil production would concomitantly drop 20 percent.

U.S. import of crude oil from African countries has been declining steadily and is expected to reduce even further. But contrary to general perceptions, U.S. imports from the Middle East not only will continue but it may even rise in the short-term. The main reason for that is type of American tight oil that is considered as light and sweet. Most U.S. refineries are designed to refine heavy and sour crude oil. It will take time before their refineries switch their feedstock to light and sweet. Even under the scenario that U.S. refineries change their configuration to processing their own domestic crude oil, Saudi Arabia will still preserve its minimum market of 1 million barrels a day in the U.S. market as a feedstock for Motiva refinery in Texas, where the Kingdom holds a 50 percent share.

Due to the surge in energy supplies and slower global growth, prices should have a downward movement in the long-term. This would particularly affect Middle Eastern oil producing countries, whose economies and national budgets are more dependent on oil revenue. But also, lower energy prices would reduce the gap between the international and domestic prices – these countries would all need less money to spend on energy subsidies. The bottom line is that markets do not expect oil prices to dip below the $70-$80 per barrel range, as a reduction in tight oil supplies could again push prices up.

U.S. LNG exports and pricing will also affect the natural gas market, and create new competition for Qatar's gas exports. U.S. refined products will eat into Middle Eastern markets, but most of all the European market. On the petrochemical side, U.S. companies will have a higher advantage in gas-based petrochemical production compares to Middle Easter petrochemical factories.

The U.S. shale boom will have broad and significant effects on global energy markets, and could conceivably reshape the geopolitical landscape. This holds particularly true for global energy security. The shift in energy flow and trade, and the changes in the trajectories of supply and demand, will create major changes in the relationships between traditional supplier and customers. Major energy producers like Saudi Arabia and Russia have already started rethinking their marketing strategies and foreign relations with potential new customers. U.S. shale gas and tight oil pose a major challenge for countries like Iraq, Iran and Mexico, who are struggling to gain back their lost production capacity and markets.

Notes

1 British Thermal Units (BTU).
2 U.S. Energy Information Administration (EIA), (Washington, DC: EIA, June 2014), http://www.eia.gov/todayinenergy/detail.cfm?id=16511.
3 The share of U.S. domestic energy supplies in 2005 was around 69%, a historically low point.
4 U.S. Energy Information Administration (EIA), *Annual Energy Outlook 2014* (Washington, DC: EIA, 2014), http://www.eia.gov/forecasts/aeo/er/pdf/tbla14.pdf.
5 Ibid.
6 BP, Energy Outlook 2010 to 2030, January 2013, http://www.bp.com/content/dam/bp/pdf/statistical-review/EnergyOutlook2030/BP_Energy_Outlook_2030_Booklet_2013.pdf.
7 Exxonmobil, Outlook For Energy 2040, 2014, http://corporate.exxonmobil.com/en/energy/energy-outlook.
8 BP, Energy Outlook 2010 to 2030, January 2013, http://www.bp.com/content/dam/bp/pdf/statistical-review/EnergyOutlook2030/BP_Energy_Outlook_2030_Booklet_2013.pdf.
9 Ibid.
10 Ibid.
11 U.S. Energy Information Agency (EIA), "Technically Recoverable Shale Oil and Shale Gas Resources: An Assessment of 137 Shale Formations in 41 Countries Outside the United States," June 13, 2013, http://www.eia.gov/analysis/studies/worldshalegas/.
12 Ibid.
13 Ibid.
14 U.S. Energy Information Administration (EIA), *Annual Energy Outlook 2014* (Washington, DC: EIA, 2014), http://www.eia.gov/forecasts/aeo/er/pdf/tbla14.pdf.
15 Internaitonal Energy Agency (IEA). "World Energy Outlook 2013." Paris: IEA, 2013.
16 U.S. Energy Information Administration (EIA), *Annual Energy Outlook 2014* (Washington, DC: EIA, 2014), http://www.eia.gov/forecasts/aeo/er/pdf/tbla14.pdf.
17 West Texas Intermediate crude oil.
18 Maugeri, Leonardo. "The Shale Oil Boom: A U.S. Phenomenon" Discussion Paper 2013-05, Belfer Center for Science and International Affairs, Harvard Kennedy School, June 2012.
19 Ibid.
20 Maugeri Leonardo, *Could we have too much oil by the end of decade?* (Netherlands: EAGE, 2012).

21 OPEC, Monthly Oil Market Report, (Vienna: OPEC, April 2014), http://www.opec.org/opec_web/static_files_project/media/downloads/publications/MOMR_April_2014.pdf.
22 OPEC, Saudi Arabia facts and figures, http://www.opec.org/opec_web/en/about_U.S./169.htm.
23 U.S. Energy Informationa Administrations (EIA), "Saudi Arabia Country Analysis Brief," http://www.eia.gov/countries/analysisbriefs/Saudi_Arabia/ saudi_arabia.pdf.
24 OPEC. *Kuwait facts and figures* . http://www.opec.org/opec_web/en/about_U.S./165.htm.
25 Ibid.
26 Sarah O. Ladislaw, Maren Leed, Molly A. Walton. *New Energy, New Geopolitics, Balancing Stability and Leverage.* Center for Strategic and Internatioal Studies (CSIS), Washington, DC.: CSIS, April 2014.
27 Maugeri, Leonardo. "The Shale Oil Boom: A U.S. Phenomenon" Discussion Paper 2013-05, Belfer Center for Science and International Affairs, Harvard Kennedy School, June 2012.
28 Kimmeridge Energy. *How will tight oil impact global oil prices this decade?* Kimmeridge Energy, 2012.
29 U.S. Energy Information Administration (EIA), *Annual Energy Outlook 2014* (Washington, DC: EIA, 2014), http://www.eia.gov/forecasts/aeo/er/pdf/tbla14.pdf.
30 Term contracts between ESSAR of Venezuela and Reliance of India.
31 Fesharaki, Fereidun. "Is the U.S. a Rising Energy Superpower?" *Center for Strategic and International Studies (CSIS), Energy Program Event.* Washington, DC., May 16, 2014.
32 Ibid.
33 Saudi Arabia has around 2 million barrels a day refinery capacity outside of its country not only in the United States but also in Japan, China, Philippines and South Korea. This refinery capacity is through the Kingdom's joint venture and investments on refineries that are design to process Saudi Arabi's crude oil as their feedstock.
34 BP, *BP Statistical Review of World Energy 2013* (London: BP, 2013), http://www.bp.com/content/dam/bp/pdf/statistical-review/statistical_review_of_world_energy_2013.pdf.
35 Liquefied Natural Gas.
36 Fesharaki, Fereidun. "Is the U.S. a Rising Energy Superpower?" *Center for Strategic and International Studies (CSIS), Energy Program Event.* Washington, DC., May 16, 2014.
37 Liquid Petroleum Gas.
38 Source of the Table: U.S. Energy Information Administration (EIA), *Analysis and Projections, Technically Recoverable Shale Oil and Shale Gas Resources* (Washington, DC: EIA, 2013), http://www.eia.gov/analysis/studies/worldshalegas/pdf/fullreport.pdf.
39 Source of the Table: Ibid.
40 Source of the Table: Kleinberg, Robert. "Shale Gas & Tight Oil Technology Evolution & Revolution ." Cambridge, Massachusetts: Shlumberger, April 2014.
41 U.S. Energy Information Administration (EIA). "Glossary of terms". http://www.eia.gov/tools/glossary/index.cfm?id=C.
42 Internaitonal Energy Agency (IEA). "World Energy Outlook 2013." Paris: IEA, 2013.
43 Ibid.
44 (EIA), U.S. Energy Information Administration. *Glossary of terms.* http://www.eia.gov/tools/glossary/index.cfm?id=C.
45 Ibid.
46 Ibid.

References

Administration, U.S. Energy Information. *Glossary of terms.* http://www.eia.gov/tools/glossary/index.cfm?id=C.
British Petroleum (BP). *Energy Outlook 2010 to 2030.* Anual Energy Outlook, London: PB, 2013.
Exxonmobil. *Outlook for Energy.* Energy Report, Exxonmobil, 2014.
Fesharaki, Fereidun. "Is the U.S. a Rising Energy Superpower?" *Center for Strategic and International Studies (CSIS), Energy Program Event.* Washington, DC., May 16, 2014.

Internaitonal Energy Agency (IEA). "World Energy Outlook 2013." Paris: IEA, 2013.
International Energy Agency (IEA). *World Energy Outlook 2013* . Annual Outlook, Paris: IEA, 2013.
Kimmeridge Energy. *How will tight oil impact global oil prices this decade?* Kimmeridge Energy, 2012.
Kleinberg, Robert. "Shale Gas & Tight Oil Technology Evolution & Revolution ." Cambridge, Massachusetts: Shlumberger, April 2014.
Maugeri, Leonardo. "Could we have too much oil by the end of the decade?" *First Break*, October 2012.
OPEC. *Kuwait facts and figures* . http://www.opec.org/opec_web/en/about_us/165.htm.
OPEC. *Monthly Oil Report.* Monthly Oil Report, Vienna: OPEC, 2014.
OPEC. *Saudi Arabia facts and figures.* http://www.opec.org/opec_web/en/about_us/169.htm.
Sarah O. Ladislaw, Maren Leed, Molly A. Walton. *New Energy, New Geopolitics, Balancing Stability and Leverage.* Center for Strategic and Internatioal Studies (CSIS), Washington, DC.: CSIS, April 2014.
U.S. Energy Infomarion Administration (EIA). *"Technically Recoverable Shale Oil and Shale Gas Resources: An Assessment of 137 Shale Formations in 41 Countries Outside the United States,"* . Assessment of Recoverable Shale Resources, Washington, DC: EIA, 2013.
U.S. Energy Infomation Administration (EIA). "Monthly Energy Review." Monthly energy report, Washington, DC, 2014.
U.S. Energy Information Administration . "Monthly Energy Review, June 2, 2014." Energy Review, Washington DC, 2014.
U.S. Energy Informationa Administrations (EIA). "Saudi Arabia Country Report."

Appendix

List of Tables

Table 1 Top 10 Countries with Technically Recoverable Tight Oil Resources[38]

Rank	Country	Tight oil (Million barrels)
1	Russia	75
2	U.S.	58
3	China	32
4	Argentina	27
5	Libya	26
6	Australia	18
7	Venezuela	13
8	Mexico	13
9	Pakistan	9
10	Canada	9
World Total 345		

Table 2 **Top 10 Countries with Technically Recoverable Shale Gas Resources**[39]

Rank	Country	Shale Gas (Trillion Cubic Feet)
1	China	1,115
2	Argentina	802
3	Algeria	707
4	U.S.	665
5	Canada	573
6	Mexico	545
7	Australia	437
8	South Africa	390
9	Russia	285
10	Brazil	245
World Total 7,299		

Table 3 **Performance of Tight Oil Well versus Conventional Oil Well**[40]

	Tight Oil Well	Conventional Oil Well
Recovery Factor	5%	50%
Annual Decline Rate	50%	5%

List of Graphs

Chart 1 **North American Tight Oil Production**

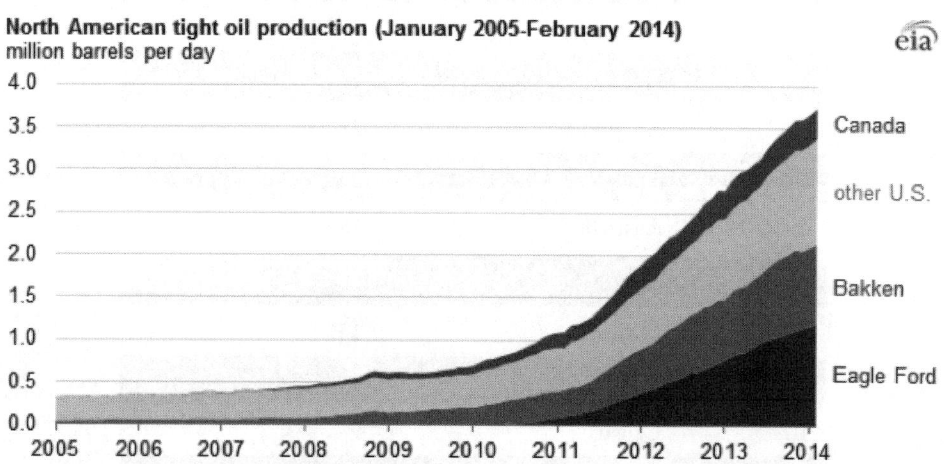

Source: U.S. Energy Information Agency (EIA)

Chart 2 Tight Oil Production in U.S.

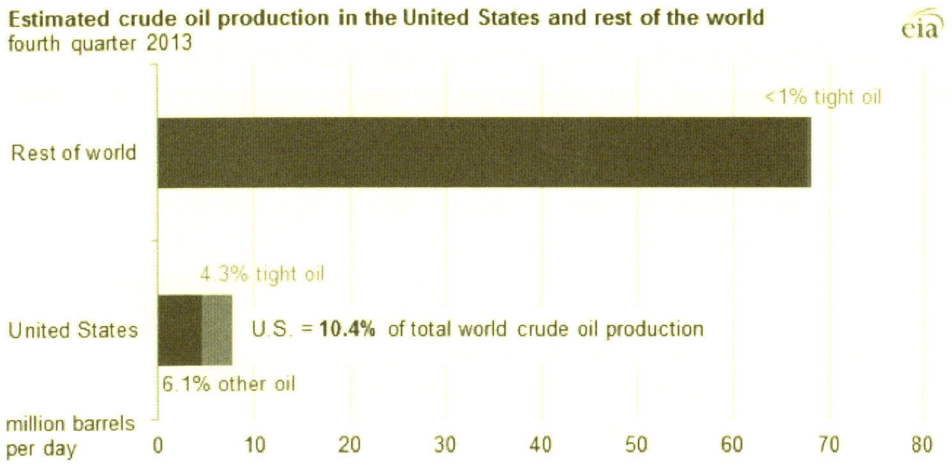

Source: U.S. *Energy Information Agency* (EIA)

Chart 3 U.S. Exports of Finished Petroleum Products (thousand barrels per day)

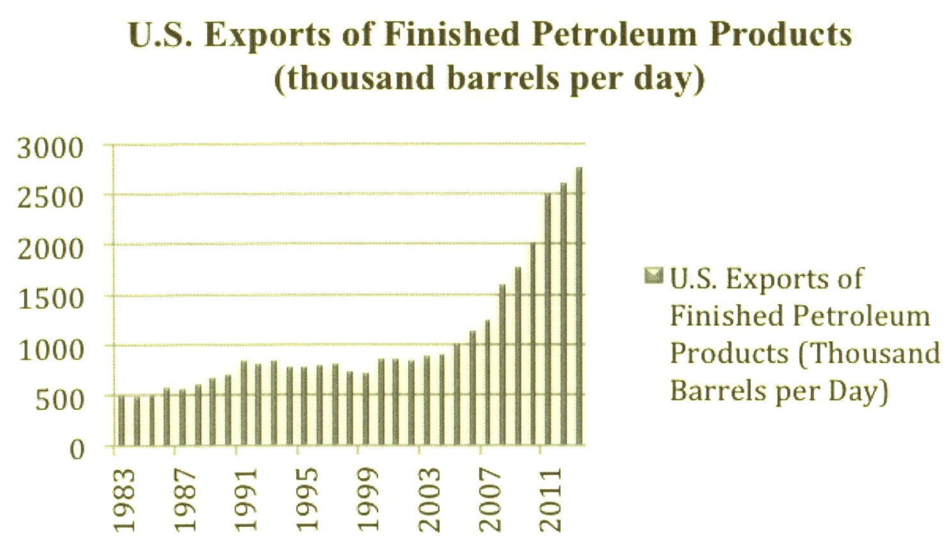

Source: U.S. Energy Information Administration

Definitions of Technical Terms that Have Been Used in this Paper

Conventional Oil and Natural Gas – According to EIA, conventional oil and natural gas will be extracted by drilling a well into a geological formation with fluid characteristics that allows the oil and natural gas to move to the wellbore.[41]

Unconventional Resources – Unconventional resources do not produced with the same methods as conventional resources. These resources could only produced economically with the help of distinct technologies of hydraulic fracturing and horizontal drilling.[42]

Unconventional Oil – According to IEA, unconventional oil includes extra-heavy oil, light tight oil, bitumen (oil sand), kerogen oil (oil shale), gas-to-liquid (GTL), coal-to-liquids (CTL) and additives.[43]

Unconventional Gas – The common method for extracting unconventional gas is hydraulic fracturing. Unconventional gas resources include tight gas, shale gas, Methane hydrates and coalbed methane (CBM).

Tight Oil – Tight oil is also know as light tight oil (LTO), is a petroleum consist of light crude oil that is contained in a low permeability shale formation. Tight oil can be produced with the help of horizontal drilling and hydraulic fracturing.

Shale Gas – Shale gas is a natural gas produces from shale formations and produced with combination of hydraulic fracturing and horizontal drilling.

Oil Sand – Oil sand that is also known as tar sand or bitumen is a type of heavy unconventional oil. Canada has the highest recoverable oil sands reserves of 800 billion barrels.

LPG – LPG is a group of hydrocarbon gases that have been liquefied at the high pressure for the purpose of storage or transportation. LPG primarily consists of propane, isobutene and normal butane produced in crude oil refinery or natural gas processing.

LNG – LNG is a natural gas (mainly methane) that has been liquefied in for the purpose of transportation by ship. Reducing the temperature to minus 260 degrees Fahrenheit will liquefy the natural gas. The transported LNG will be re-gasified again by increasing the temperature at the gasification units at the destination.

Naphtha – Naphtha is a refined or partially refined light distillate that could be used with mix of other chemicals as high-grade gasoline or jet fuel. It could also be used as a feedstock for petrochemical factories. The boiling point range of naphtha is between 122 and 400 degrees Fahrenheit.[44]

Hydraulic Fracturing – A technology of extracting liquids from low permeability formations with injecting large volumes of liquids (water mixed, chemicals and sand) with a high pressure to create cracks in the rock. The cracks let the tapped gas or liquids to free and flow into the wellbore.[45]

Horizontal Drilling – Horizontal drilling is a process of drilling a non-vertical well, which allows access to higher surface to the well.[46]

10

Converting the Finite to Infinite: A Comparative Study for Qatar and the United Arab Emirates (UAE)

Dania Thafer

Background

There is a fundamental concern among the Gulf Cooperation Council Countries (GCC) and the broader Middle East about the sustainability of the current economic model. As energy demand increases and the population continues to grow, it is questionable whether oil supplies can sustain domestic energy needs while simultaneously maintain profitability from its hydrocarbon exports in the long-run. "Energy consumption across the region has nearly tripled since 1990 and grew by 72% between 2000 and 2010 – about three times faster than the rate of growth across the world as a whole, driven by a growing population and the rise of energy-intensive industry – not least oil and gas extraction. This phenomenal growth of energy consumption is outstripping that of energy production across the region.[1]"

Within the Middle East, the GCC countries, even with their small populations have been projected to have considerable oil and gas demand growth of more than 50 percent by 2030.[2] With this growth, there are three significant issues for their energy efficiency (1) the high level of fuel subsidies; (2) the process of conversion to renewable energies; (3) and the cost and sustainability of their water sources through desalination. Renewable energy solutions are also important in addressing their astronomically high carbon emissions. The GCC only consists of 0.6 percent of the world's population, yet 2.4 percent of the world's emissions stem from there.[3] Of the smaller oil wealthy Gulf countries, Qatar and UAE make interesting case studies for launching energy mix initiatives.

In both the UAE and Qatar, the fuel prices are less than half of market prices with a subsidization rate of 64.1 percent and 75.1 percent, respectively.[4] Not only is domestic demand increases a result of a growing population, but maintaining artificially low fuel prices through government subsidies can be considered a major contributing factor to the increased domestic demand of hydrocarbon resources. Along with the other GCC countries, Qatar and the UAE increasingly rely on their hydrocarbon production to meet domestic demands. Both are among the handful of countries worldwide that keep gas prices artificially low and are substantial contributors to the more than $500 billion spent annually on subsidies worldwide.[5] A study implemented by the OECD stated that if subsidies were removed worldwide that carbon emissions can be lowered by more than 10 percent by 2050.[6] Although there would be a clear benefit in curbing the increased domestic demand, there are other consequences that could result of removing fuel subsidies—such as a looming prospect of a hefty political price. For example, when Jordan raised its fuel prices there were angry crowds protesting violently in the streets. In 2010 the UAE raised fuel prices, but shortly thereafter in 2011 the government decided against raising it further wary of stoking discontent similar to the popular uprisings in the region. Qatar has not shown any tangible signs of interest in reducing its subsidies. Although both the UAE and Qatar have minimally progressed with regards to subsidies, they have introduced renewable energy strategies to their energy mix – especially with regards to solar energy.

One resource that the Gulf countries have in abundance is solar energy. According to the Solar GCC Alliance, within the last decade investment in solar energy has skyrocketed from $5 billion in 2003 to $93 billion in 2014.[7] Also, wind power is another renewable energy that is being explored among the Gulf countries. With the dramatic increase in electricity and gas demand in Qatar and the UAE, there will need to be a significant increase in power-generating capacity in the upcoming years. Solar and wind power could be effective complements to the existing conventional capacity and provide electricity about half the cost of market prices of oil-fired power generation. Both the UAE and Qatar reveal robust development in solar and some initiatives in wind power. An area of great opportunity for the water-scarce Gulf countries is utilizing renewable energy in the water desalination process.

Among the GCC countries one of the most critical challenges is having a sustainable approach to securing water resources. The GCC countries use desalination as a strategic water resource and produces 39 percent of the world's desalinated water.[8] The UAE is the second largest producer of desalinated water followed by Saudi Arabia. Both Qatar and the UAE have announced their interest and plans to use solar and wind energy to produce desalination technologies. Regardless of the amount of interest shown among the UAE and Qatar, the application of solar and wind power to drive desalination plants is still minimal and has plenty of room for growth.

UAE Energy Mix

Demand Destruction & Fuel Subsidies

There have been ongoing debates in the United Arab Emirates on whether there should be a decrease or increase in energy subsidies. The UAE produced in 2013 roughly 3.65 mbd of oil and its oil exports account for about half of the GDP.[9] The issue of oil consumption and subsidies is a topic of sensitivity due to the stark increase in domestic demand and it being a growing burden to continue meeting the population's expectation of subsidized energy. In the last decade there has been a substantial increase in domestic consumption; in 2003 domestic consumption in the UAE was at 453,000 b/d and by 2013 it was 773,000 b/d which is a 70.6 percent growth, while production grew only 34 percent (from 2.72 MBD in 2003).[10] This reveals that if the consumption continues on this trajectory, it could continue to catch up to production levels and adversely affect oil revenues. For example, in 2003 16.6 percent of oil production was consumed domestically and by 2013 it increased approximately 5 percent to 21.2 percent[11]. Along with these concerns, is the fact that the hydrocarbon-wealthy neighboring countries have relatively higher subsidies for their citizens, leaving a disenchanted UAE population. According to the International Energy Agency, in 2012 the UAE had an average subsidization rate of 64.1 percent; whereas its counterpart Kuwait's rate is 96 percent and Qatar at 75.1 percent.[12] Although the UAE is a major oil producer in 2012, it had the third-highest fuel prices in the Arab world after Tunisia and Syria.

In the UAE, the federal government sets fuel prices and gasoline sells at a much lower price than the original market price. The UAE government subsidizes 1.20 dirhams per liter of standard grade gasoline which was sold at 1.72 dirham (or $0.47) per liter in 2012.[13] The UAE government did increase fuel prices 0.35 dirhams in 2010 in an attempt to mitigate the economic burden and boost efficiency. Even with the price increase in 2010, the UAE still spent 5.5 percent of its gross domestic product on fuel subsidies.[14] After the events of the Arab Spring, plans to further institute higher gasoline prices by decreasing subsidies were postponed in order to mitigate any potential discontent among the population.

The Federal National Council and industry advocates are in disagreement about the implementation of gasoline subsidies. As a response to UAE nationals' complaints, the UAE's Federal National Council unanimously approved a recommendation to cut gasoline prices for all residents in 2012. They were advocating for prices that will match their GCC counterparts and they argued that being one of the largest oil-producing countries, they should not have the third-highest oil prices among Arab countries. The FNC committee further argued that limited refining capacity was a key factor in which resulted in high import needs and hence the losses for retailers. However, industry analysts disagree. They accredit the problem to low fuel prices due to subsidies and stated that these subsidies discourage the necessary investment due to lack of industry profitability.

In addition to other issues, there is an imbalance among gasoline retailers between the emirates since the majority of oil is produced in Abu Dhabi and the emirates are facing challenges absorbing the losses from subsidies. Just the combined losses of state-owned retailers alone were 8.5 billion in 2011 and are expected to increase.[15] For example, the shutdown in Sharjah and Ras Al-Khaimah of ENOC along with its subsidiary EPPCO forced Emiratis from the north to commute long distances to fill their tanks. In addition, Dubai's government has shown an unwillingness to independently take responsibility of subsidizing consumption of residents of other emirates by declining to serve them. Unlike Abu Dhabi, Dubai-based retailers have to import oil as a loss, making oil consumption a hefty burden. Abu Dhabi has taken initiative to mitigate the pressures on other less-wealthy emirates by ADNOC signing an agreement to take over 74 service stations in Fujairah, Ajman, Sharjah and Ras Al Khaimah, and Umm Al Quwain in 2012.[16] Given that the UAE is one of the top-five power consumers per capita, the IMF has issued a warning that the UAE should reduce energy subsidies to relieve some of Dubai's debt and mitigate the risk of absorbing increased prices.[17]

Solar Energy

"Worldwide since 2004, global concentrated solar plant (CSP) capacity has increased 10-fold, and last year surged 36% to a total of 3.4 gigawatts of energy generated."[18] In the Gulf solar power is a renewable energy source that is naturally abundant. Abu Dhabi alone suffers from having the second highest carbon dioxide emissions worldwide per capita each day. To resolve some of these issues, the UAE is targeting to achieve a 7 percent renewable energy capacity by 2020, while Dubai's Integrated Energy Strategy aims to attain a 5 per cent solar consumption of total power consumed.[19] The UAE has a number of solar energy projects that are emerging to reduce dependence on hydrocarbon resources. For example, Dubai Electricity and Water Authority (DEWA) is in its second phase of establishing the Mohammad Bin Rashid Al Maktoum Solar Park, which consists of 1,000MW.[20]

The UAE is ranked third worldwide in 2013 for its CSP capacity and its CSP technology investment. According to REN21, an international network established for advocating for renewable energy, the UAE is a major influencer for emerging markets' tripled CSP growth in 2013 due to the 100 megwatt CSP plant in western Abu Dhabi. The $600 million Shams 1 initiative, the largest renewable energy project in the Middle East under the Masdar initiative, can be credited for these accomplishments. Since its launch it diminished the release of 175,000 tons of carbon annually and generated enough powers to offer electricity to 20,000 homes in UAE.[21]

Among large-scale CSP plants, Shams 1 stands out for its innovative approach towards water consumption and conservation. Typical CSP plants consume vast quantities of water because of the need to cool and condense steam before it is pumped back into the generator. Shams 1 instead uses an air-cooled condensing system, which saves hundreds of millions of gallons of water annually while maintaining an overall high operational efficiency.[22]

Wind Power

Wind power is another renewable source that is not as abundant among the GCC countries as solar energy. In order to identify a viable location for wind power, the UAE is launching a wind-mapping project to identify the most ideal locations to place turbines. Sir Bani Yas Island is one location in which Masdar proposed a 30MW project. Securing a location that is prone to efficient wind power requires the assessment of many variables in the UAE. Although recent developments reveal that power can be generated at much lower velocities, one factor that makes it challenging to apply this particular renewable energy strategy is the relatively weak wind speeds in the UAE. Furthermore, another issue to consider regardless of how susceptible the mountainous region of the northern Emirates is to wind velocities, the mountain requires additional logistical and infrastructure considerations in building turbines.[23]

Desalination & Renewable Energy

The Gulf countries have ample amount renewable energy opportunities that should be tapped into. The MENA region has more than 16,5000MW of renewable energy capacity in which the majority can be derived from hydropower.[24] One area of concern for the GCC countries is maintaining ample water supplies. "According to the UN, over 40 per cent of the global population is affected by water scarcity, and by 2025, 1.8 billion people will live ... (with) ... water scarcity."[25] To maintain water supplies, the GCC countries use the desalination process to convert salt water into usable water and this process utilizes a high level of energy. Approximately 52 percent of desalinated water produced is from the Gulf countries.[26] Applying renewable energy processes to water desalination is one approach to decreasing high energy consumption.

Desalination is a very important process for the water-scarce GCC countries in which seawater and brackish water is converted to freshwater supplies. The drawbacks of this process is that it is very expensive and energy intensive. A major issue of concern is that it is highly susceptible to the logistical constraints and volatile global market prices of fossil fuel sources. Renewable energy powered desalination could be a key strategy in reducing the burdensome cost of maintaining an ongoing supply of freshwater. Although currently 1 percent of water desalination utilizes renewable energy, there is ample opportunity for increases in this area. "For example, solar energy – in particular heat from CSP for thermal desalination and electricity from solar photovoltaic and CSP for membrane desalination – is a key solution in arid regions (e.g. the MENA region) with extensive solar energy potentials, whilst wind energy is of interest for membrane desalination projects in coastal and is-lands communities."[27]

"For UAE, the cumulative water production increased from 6.7 Mm 3/day (in 2006) to 17.5 Mm3/day (in 2025). This large increase in water production will be associated with a similar large increase in energy requirement. The results show that the annual total energy consumption in 2006 was 64,762 GWhr and will almost triple by the year 2025 to be 145,412 GWhr. Thirty small scale desalination water production (50 m3/day) using solar PV - BWRO plants are built in UAE with ambitious plan to build 100 MIGD RE driven desalination plants by 2020. No thermal solar desalination plants are built since the decommissioning of UAE Umm El- Naar Solar-MED plant. Renewable and possibly nuclear energy have been announced by several Gulf States to partially operate new desalination projects."[28]

The UAE derives 30 percent of its water supply from the process of desalination. One of the goals of the UAE is to desalinate water with renewable energy. Water desalination is one of the major contributors of carbon dioxide emission and Abu Dhabi has the second highest emission worldwide.[29] In order to increase efficiency and reduce carbon dioxide emissions, Masdar was hired with an aim to have all desalination plants powered by renewable energy. In May of 2014, Masdar along with the Environment Agency Abu Dhabi (EAD) and the Abu Dhabi Water and Electricity Company awarded four international renewable energy companies. "These are Abengoa (Spain), working with a capacity of 1,080 cubic metres of water per -day for this project, Degremont (France), with 100 cubic metres, Sidem (France) with 300 cubic metres and Trevi Systems (USA) with 59 cubic meters."[30] These companies have been hired to implement a trial session of a renewable energy desalination plant at the border between Dubai and Abu Dhabi. They have not disclosed the renewable energy mix in which they will utilize; they are still trying to strike the ideal balance between technology and cost. This project is intended to launch for a trial between January 2015 and June 2016.

Qatar Renewable Energy Mix

Demand Destruction & Fuel Subsidies

Qatar has begun creating a strategy and campaign to enhance energy efficiency in the country. Both the Qatar National Development Strategy 2011-2016 and Qatar's Vision 2030 are targeting to release awareness campaigns for efficiency, implementation of standardization and apply seasonal shutdowns. There is no explicit target of reduction, however, it has been stated that this strategy can decrease total power generation by the year 2016 by total of 7 percent.[31] Although Qatar started initiatives for energy efficiency, one major concern is the high fuel subsidies. It has one of the highest average annual subsidy amount worldwide with a cost $2622 spent per capita and a CO2 emission rate of 50 tons per capita annually.[32] Its CO2 emissions are 10 times more than the world income loss from fuel subsidies, ranking #9 at approximately $300 million average of 4.6 tons per capita.[33] Furthermore, in 2010 subsidies accounted for 15.39 percent of government expenditures[34] and in 2012 it managed to be one of the top 10 worldwide with deadweight costs (pg. 16). When inquired about the reduction of energy subsidies in Qatar, the Minister of Energy, HE Dr. Mohammed bin Saleh Al Sada argued in December of 2012 that the notion of cheap or expensive is subjective, and subsidies are closely linked to the economic health of the country; therefore, he purported that a close evaluation of the consequences of a subsidy reduction needs to be conducted before any changes are executed.[35] On the other hand, "Qatar's National Development Strategy 2011-2016 states it more bluntly, saying fuel subsides are "at odds with the aspirations" and sustainability objectives of the wealthy emirate."[36]

While Qatar has an astronomical amount of gas capacity, its oil capacity is limited relatively speaking. "Qatar managed to produce 77 million tonnes per annum (MTa) of liquefied natural gas (LNG). The country's North Field holds the largest concentration of natural gas on earth, at 900 trillion cubic feet - 15 per cent of the global total. In contrast,

Qatar holds modest oil reserves of around 15 billion barrels, providing a daily output of 800,000 barrels. Projections show that current oil reserves should ensure continued output for the next 23 years."[37]In the last decade there has been an increase in domestic oil consumption; in 2003 domestic consumption in Qatar was at 95,000 b/d and by 2013 it was 267,000 b/d which is a 181 percent growth, while production grew 110 percent (from 949,000 b/d in 2003 to 1,995,000 b/d in 2013).[38] This reveals that if the consumption continues on this trajectory, it could continue in the long-term to catch up to production levels and adversely affect revenues. For example, in 2003 10 percent of oil production was consumed domestically and by 2013 it increased approximately 3 percent to 13.4 percent[39]. "Qatar has one of the highest energy consumption per capita in the world with nearly 50-70% of the energy consumption used for cooling, and 15-20% is for lighting."[40]

It is important that Qatar creates a policy that curb high demand levels in order to ensure revenues remain intact and that carbon emissions are at more reasonable rates. One approach to decreasing domestic demand of hydrocarbon energy is to invest more in renewable energy.

Solar Energy

With an average duration of 9.4 hours daily sunshine, solar energy is the most significant form of renewable energy potential and Qatar has taken large strides in establishing a strategy to tap into this renewable energy.[41]"According to estimates from QEERI, a 250-square-kilometre site in Qatar could support 10,000MW of solar power capacity."[42] In December of 2013, Dr. Mohammed Al Sada, the Energy Minister, announced plans that Qatar will derive 2 percent (or 200 MW) of their power from renewable sources by 2020. Within the next six years, Qatar is also targeting to generate 200MW in which is expected to provide enough energy to power up to 66,000 homes annually.[43] Furthermore, the largest solar energy producer in the country, Qatar Foundation, released a statement that it plans to increase its production by approximately 150 percent or an additional 5MW in the upcoming years. This is in addition to its current 3.3MW production level.[44] [45]

There are several ongoing and emerging solar energy projects that are being piloted by Qatar. Qatar Foundation is responsible for the majority of Qatar's total solar which accounts for 85 percent the country's photovoltaic power.[46] According to Ibrahim Al-Haidos, project manager for Qatar Foundation's Solar Smart Grid, the foundation currently produces enough energy to power 471 Qatari houses.[47] Part of the foundation's new initiative is to monitor solar power generation and they recently launched a new Energy Monitoring Center in which is unique to the GCC and will be monitoring the foundation's solar power generation. In addition to the foundation, Qatar Solar Energy opened the first 300 MW PV module production plant in the Middle East."[48] Qatar Solar Energy board member Kamel Ounadjela said 300mw would provide enough energy for 100,000 homes each year in high energy-consuming countries such as Qatar or the US."[49] Also, Qatar Solar Energy stated it will begin a three-phase plan to produce 2.5GW of solar energy annually. Along with these initiatives, Kahramaa, a state-run electricity and water company, is launching a $34 million 5MW-10MW project.[50] To further its technological advancement in the field, Qatar will be hosting a Solar Qatar summit that will attract 300 experts of the field. Industry analysts have projected that the 2022 World Cup plan of instituting a solar powered stadium cooling system will create the perfect storm for an emergence of strong solar industry in the country. ". It is proposed that solar energy systems will harness the desert sun and use it to power the stadiums' cooling systems. The solar energy installations will be connected to the stadiums as well as the national grid. Generators using biofuels will supplement electricity from the national grid to ensure that there is enough capacity to cool the stadium during matches. Surplus energy will be exported back on to the grid. The amount of electricity generated is intended to exceed the amount of electricity imported for events over the year, making the facility carbon neutral."[51]

Wind Power

Given the minimal wind power capacity relative to solar energy, wind power is an area of peripheral focus. Comparatively speaking there is minimal research and effort in the topic. Furthermore, there are no wind atlases or large-scale wind projects. "The average wind speed was about 4.4 metres a second, with a monthly peak of 5.5 metres a second being recorded in June and a low of 3.6 metres a second in September."[52] These wind speeds can support small to medium wind turbines.[53] Some argue that these winds can support endeavors such as water pumping and providing electricity to remote locations off grid.[54] Although Qatar has limited wind potential, it does have more promising irradiation levels inland that are high enough to support power generation.

With its minimal wind capacity, there are still a couple of wind power-related projects in Qatar. In 2011, Vodafone Qatar Q.S.C. and Alcatel-Lucent announced that they will launch the first hybrid powered Base Station with both solar and wind energy in Qatar. The 1kW station was powered by with a 2.5kW turbine and 3kW solar array providing 90 per cent of its power and has been carrying traffic since 2010.[55] Also, Qatar's sovereign wealth fund, Qatar Investment Authority, has taken steps in investing in the renewable energy sector. In March 2011, the Qatar Investment Authority acquired a $2.6 billion or percent stake of Iberdrola, a Spanish company that focuses on windpower, in order that Qatar can access its cleantech operations.[56]

Desalination & Renewable Energy

Qatar has experienced a soar in demand for water and electricity due to its substantial population growth and strong commercial expansion. A key area of concern is maintaining an adequate and sustainable water supply. It is projected that the cumulative water production increase substantially from its 2006 rate of 1.2 Mm 3/day in 2006 to 3.2 Mm 3/day by 2025, more than doubling.[57] This large increase in water production will naturally be associated with a comparable increase in energy requirements since the process of deriving the water sources require a substantial amount of energy consumption. This is where renewable energy can play a strong role in mitigating against the depletion of hydrocarbon sources and government resources. In addition, food security and agriculture is linked to the availability of ample water sources.

In 2008 Qatar embarked on the Qatar National Food Security Program (QNFSP) in which its mission is to develop domestic agriculture production to improve food security. The program aims to reverse Qatar's 90 percent importation of food which is a significant security risk and since 2008 there was a 30 percent decline in local food production. "The National Plan that envisages to augment Qatar's current domestic production of agriculture from the existing capacity of 8 percent against the country's total demand to 40-60 percent in short term to mid-term."[58] With Qatar suffering from depleting natural water reserves, the water for agricultural endeavors must come from sea water desalination. Therefore, taking into consideration the fact that water desalination is highly intensive in energy consumption and its negative consequences for the environment, QNFSP has sought out to utilize combined solar energy power for water desalination. The water demand for the project was between 100 million g/d and 296 million g/d which requires solar power.[59]« Through developing vast tracts of desert using solar-powered desalinated water in Qatar and acquiring agricultural land overseas, the government is aiming to limit imports from third parties to 30 per cent of all food needs by 2023.[60]

Conclusion and Recommendations

Creating an energy mix that is conducive to the demands, resource constraints and long-term vision of a country requires substantial investment and a deliberate strategic outlook. It is necessary to ensure efficiency by managing demand levels incentivizing efficiency and subsidies is one approach that distorts energy demand to unfavorable results. Not only applying strategies to curb demand is vital, but also establishing an energy mix that promotes sustainability through renewable energy.

Table 1 **2012 Fossil Fuel Subsidy Rate in GCC**

Saudi Arabia	78.9%
Kuwait	96%
United Arab Emirates	64.1%
Bahrain	39.5%
Qatar	75.1%

Source: International Energy Agency. "Fossil Fuel Consumption as a Proportion of Full Cost of Supply, 2012". http://www.iea.org/subsidy/index.html (accessed July 1, 2014).

In the Middle East, socio-economic policies in regards to demand destruction have been counter the interest of increasing efficiency and promoting diversification in energy sources. These policies have made fuel and electricity prices artificially low at a rate that is one-third of the world's average price; where worldwide a liter of fuel costs $1.90 versus $50 per liter in the Middle East.[61] The UAE and Qatar are no exception. Their citizens only pay 35.9 percent and 24.9 percent of the cost of fuel on average, respectively (see Table #1 2012 Fossil Fuel Rate in the GCC). In the UAE, there has been tension between industry and the Federal National Council about whether subsidies should remain intact. The fact that the UAE has the lowest subsidies compared to its hydrocarbon-wealthy counterparts, coupled with the fact that it has decreased its subsidies by .35 dirhams in 2010, reveal that the UAE has put an exerted effort to curb domestic demand. However, these initiatives came to a halt in 2011 when unrest swept the region. On the other hand, although Qatar's National Development Strategy 2011-2016 stated that fuel subsides are "at odds with the aspirations" and sustainability objectives of the wealthy emirate, the Minister of Energy, HE Dr. Mohammed bin Saleh Al Sada mentioned hesitantly in 2012 that decreasing subsidies is a sensitive issue that is subject to review. On the front of subsidization, the UAE has taken more tangible initiatives in decreasing subsidies. However, the situation seems as it is more a matter of survival since Abu Dhabi is the only truly oil-wealthy emirate of the UAE; whereas Qatar has an ample amount of hydrocarbon resources, especially when considering its size.

Both countries should evaluate a more aggressive approach to subsidization, given citizens are generally wealthier per capita and have other social benefits that can outweigh higher fuel costs. Also, since hydrocarbon resource revenues are the lifeline of these countries, preserving their reserves are extremely important to the economic health of the country. Generally speaking, energy subsidies are an important social safety-net for targeting the poor or less fortunate, but in both Qatar and UAE the citizens generally live well above the poverty-line and this may be a policy that can be reevaluated if poverty is not an area of concern. Also, for the poor there can be calculated strategy that can utilize targeted cash transfers or a more targeted strategy.[62] Another targeted approach is to consider letting companies and expatriates pay the full-price for fuel, while subsidies can remain a social benefit for citizens.

Competitiveness and sustainability is a major area of concern since resources used to subsidize energy cannot be invested in initiatives that promote economic growth. It is palpable that there can be economic consequences of increasing fuel prices, but reduced energy subsidies allow for energy diversification and competitiveness promotion across the economy which is vital to a healthy economy. "Energy subsidies distort price signals, with serious implications on efficiency and the optimal allocation of resources. Energy subsidies also tend to be regressive, with high-income households and industries proportionately benefiting most from low energy prices."[63]

After reviewing the ramifications of subsidies on efficiency and the economy, most industry stakeholders could agree that it is recommended to have a phased-approach to decreasing subsidies.[64 65] The UAE intended on a phased approach but the regional events of 2011 brought it to a halt. Before implementing a strategy, there should be a comprehensive energy sector reform plan that includes a study of policy implications; clearly articulates the objectives of the plan; and communicates the means of how to realize these objectives. Along with a reform plan, a campaign should be launched revealing the country-wide benefits of decreasing subsidies to hedge against population discontent. Furthermore, within the reform plan, strong institutional reforms need to be an area of focus so that as time progresses, Qatar and the UAE can sustain a fluid and evolving strategy in managing policies relevant to subsidies. With a strong institutional base, energy pricing should be depoliticized and automatic pricing mechanisms should be introduced.

Subsidies are only one aspect of an efficient energy policy. As part of a reformation strategy, both the UAE and Qatar should capitalize on infinite renewable energy options available in order to preserve their finite hydrocarbon resources. This will allow for their limited hydrocarbon resources to remain a source of income that the health of their economies are currently dependent on. One major benefit of renewable energy is that it offers opportunities for further economic diversification and job creation for both the UAE and Qatari economies which are still highly dependent on fossil fuels.[66] Large solar energy potential exists for both countries. They both have announced plans to increase their renewable energy capacity by 2020. As mentioned in previous sections, Qatar is forecasting to derive 2 percent (or 200 MW) of their power from renewable sources by 2020, while the UAE is targeting to achieve a 7 percent renewable energy capacity by then (Dubai's Integrated Energy Strategy aims to attain a 5 percent).

The UAE has already capitalized on some of their solar potential and is assessing the prospects of wind power. Given the fact that the UAE is ranked third worldwide in 2013 for its CSP capacity and its CSP technology investment, coupled with the fact it is a major influencer for emerging markets by tripling CSP growth in 2013, shows that the UAE is the pioneering GCC country in renewable energy development. Even though Qatar has revealed a strong intention of developing renewable energy capacity, many of these initiatives are still in their nascent stages and have yet to come into fruition. But with the World Cup awarded to Qatar in 2022, there is projected to be promising developments in the renewable energy sector in order to support a solar-powered stadium. One unique contribution from Qatar to the renewable energy field is Qatar Solar Energy's first 300 MW PV module production plant. It is the first of its kind in the Middle East.

Qatar is a unique case with regards to renewable energy for a number of reasons. It being the world's leading exporter of LNG, diversification of energy sources is not as critical of a concern as it is in the UAE. In terms of feedstock, Qatar has far less to worry about than the rest of the GCC given its ownership of the North field gas reservoir. Even though Qatari authorities realize the pertinence of energy preservation, it is not an urgent matter as it is for the UAE with regards to resource depletion. Another area of concern is the minimal existence of legislation governing renewable energy policies.[67] In addition, Qatar does not have a dedicated high-profile champion of renewable energy as prominent as Abu Dhabi's Masdar initiative. Nevertheless, despite this and its substantial energy wealth, Qatar's government has demonstrated a drive and a commitment to diversifying its economy as well as promoting sustainability. They have invested heavily in research and have an array of government-funded projects in the workings.

A differentiating factor in the strategy between Qatar and the UAE is the varying model approaches for generating electricity. Unlike the UAE and other GCC counterparts, Qatar is applying a decentralized model which generates electricity from several smaller sources rather the centralized approach the UAE has taken.[68] An example of UAE's centralized approach is the opening of the largest solar power plant for Maktoum's Solar Park. It has already began to develop its second phase that is targeted to increase capacity from 13 megawatts to a whopping 113 megawatts with an ultimate goal of producing 1000MW by 2030.[69]

In conclusion regardless of their reserve sizes, both the UAE and Qatar would be better off if they sold their hydrocarbon resources for profit, rather than burning it at home and selling it at a subsidized rate to consumers. In addition, in the long-run hydrocarbon resources are finite and eventually will deplete leaving no choice but to be import-dependent, or ideally harvest the infinite sources of renewable energy. Investing in renewable energy will assist in diversifying the economy; keeping the country immune from price volatility of the domestic fuel market; hedging against the depletion of water sources and developing a food security through water sustainability.

Notes

1 "Energy subsidies hamper Middle East." Centre for Economics and Business Research. http://www.cebr.com/reports/energy-subsidies-hamper-middle-east/ (accessed July 2, 2014).

2 "GCC energy subsidies too high, ministers say." Latest News, Pictures, Video, Multimedia. http://m.gulfnews.com/business/oil-gas/gcc-energy-subsidies-too-high-ministers-say-1.1253614 (July 2, 2014).

3 Rao, Govinda, and May Al-Kuwari . "Assessment of Solar and Wind Energy Potential in Qatar." *02nd Joint Qatar-Japan Environment Symposium.* https://www.jccp.or.jp/international/conference/docs/14assessment-of-solar-and-wind-energy-potential-in.pdf. (accessed July 1, 2014).

4 International Energy Agency. "Fossil Fuel Consumption as a Proportion of Full Cost of Supply, 2012." http://www.iea.org/subsidy/index.html (accessed July 1, 2014).

5 The Christian Science Monitor. "Fuel subsidies get scrutiny at Doha talks." The Christian Science Monitor. http://www.csmonitor.com/Environment/2012/1203/Fuel-subsidies-get-scrutiny-at-Doha-talks (accessed July 2, 2014).

6 The Christian Science Monitor. "Fuel subsidies get scrutiny at Doha talks." The Christian Science Monitor. http://www.csmonitor.com/Environment/2012/1203/Fuel-subsidies-get-scrutiny-at-Doha-talks (accessed July 2, 2014).

7 "Qatar sheds light on its solar-power future." - Middle East. http://www.aljazeera.com/news/middleeast/2014/06/qatar-solar-power-future-201468121043472545.html (accessed July 2, 2014).

8 Fath, Hassan, Ashraf Sadik, and Toufic Mezher. "Present and Future Trend in the Production and Energy Consumption of Desalinated Water in GCC Countries." *International Journal of Thermal & Environmental Engineering* 5: 155 - 1 65. (accessed July 1, 2014).

9 British Petroleum. "BP Statistical Review of World Energy June 2014." http://www.bp.com/content/dam/bp/pdf/Energy-economics/statistical-review-2014/BP-statistical-review-of-world-energy-2014-full-report.pdf (accessed July 1, 2014).

10 British Petroleum. "BP Statistical Review of World Energy June 2014." http://www.bp.com/content/dam/bp/pdf/Energy-economics/statistical-review-2014/BP-statistical-review-of-world-energy-2014-full-report.pdf (accessed July 1, 2014).

11 Production: 2003 – 2722
Consumption: 2003 – 453
16.6% of Production is consumed domestically
Production: 2013 – 3646 (grew 34%)
Consumption: 2013 – 773 (grew 70.6%)
21.2% of production is consumed domestically

12 International Energy Agency. "Fossil Fuel Consumption as a Proportion of Full Cost of Supply, 2012." http://www.iea.org/subsidy/index.html (accessed July 1, 2014).

13 Bakr, Amena. "Cheap UAE gasoline: citizens and industry square up." Reuters. http://www.reuters.com/article/2012/05/30/us-emirates-petrol-idUSBRE84T0VW20120530 (accessed July 2, 2014).

14 "U.A.E. Oil Minister: Cut Energy Subsidies to Lower Domestic Consumption - Middle East Real Time - WSJ." Middle East Real Time RSS. http://blogs.wsj.com/middleeast/2014/01/21/u-a-e-oil-minister-cut-energy-subsidies-to-lower-domestic-consumption/ (accessed July 2, 2014).

15 Bakr, Amena. "Cheap UAE gasoline: citizens and industry square up." Reuters. http://www.reuters.com/article/2012/05/30/us-emirates-petrol-idUSBRE84T0VW20120530 (accessed July 2, 2014).

16 Bakr, Amena. "Cheap UAE gasoline: citizens and industry square up." Reuters. http://www.reuters.com/article/2012/05/30/us-emirates-petrol-idUSBRE84T0VW20120530 (accessed July 2, 2014).

17 "U.A.E. Oil Minister: Cut Energy Subsidies to Lower Domestic Consumption - Middle East Real Time - WSJ." Middle East Real Time RSS. http://blogs.wsj.com/middleeast/2014/01/21/u-a-e-oil-minister-cut-energy-subsidies-to-lower-domestic-consumption/ (accessed July 2, 2014).

18 "U.A.E. ranks 3rd among world nations in total concentrated solar power capacity." UAE. http://www.uaeinteract.com/docs/U.A.E._ranks_3rd_among_world_nations_in_total_concentrated_solar_power_capacity/62109.htm (accessed July 2, 2014).

19 "EnPark, Mesia support solar industry in Dubai." UAE. http://www.uaeinteract.com/docs/EnPark,_Mesia_support_solar_industry_in_Dubai/61722.htm (accessed July 2, 2014).

20 "EnPark, Mesia support solar industry in Dubai." UAE. http://www.uaeinteract.com/docs/EnPark,_Mesia_support_solar_industry_in_Dubai/61722.htm (accessed July 2, 2014).

21 "U.A.E. ranks 3rd among world nations in total concentrated solar power capacity." UAE. http://www.uaeinteract.com/docs/U.A.E._ranks_3rd_among_world_nations_in_total_concentrated_solar_power_capacity/62109.htm (accessed July 2, 2014).

22 "U.A.E. ranks 3rd among world nations in total concentrated solar power capacity." UAE. http://www.uaeinteract.com/docs/U.A.E._ranks_3rd_among_world_nations_in_total_concentrated_solar_power_capacity/62109.htm (accessed July 2, 2014).

23 "U.A.E. looks to tap vast potential of wind power." UAE. http://www.uaeinteract.com/docs/U.A.E._looks_to_tap_vast_potential_of_wind_power/62304.htm (accessed July 2, 2014).

24 "Dubai in Top 3 Middle East Growth Markets for Solar Industry – Enpark & Mesia Supporting the Solar Industry. http://www.enpark.ae/media/press-releases/237-dubai-in-top-3-middle-east-growth-markets-for-solar-industry-enpark-a-mesia-supporting-the-solar-industry (accessed July 2, 2014).

25 Khaleej Times. " Desalination plants to run on renewable energy in Abu Dhabi." - Khaleej Times. http://www.khaleejtimes.com/kt-article-display-1.asp?xfile=data/nationgeneral/2014/May/nationgeneral_May48.xml§ion=nationgeneral (accessed July 2, 2014).

26 "Desalination plants to run on renewable energy in Abu Dhabi." UAE. http://www.uaeinteract.com/docs/Desalination_plants_to_run_on_renewable_energy_in_Abu_Dhabi/61530.htm (accessed July 2, 2014).

27 "Water Desalination Using Renewable Energy: Technology Brief." *International Renewable Energy Agency* http://www.irena.org/DocumentDownloads/Publications/IRENA-ETSAP%20Tech%20Brief%20I12%20Water-Desalination.pdf. (accessed July 1, 2014).

28 Fath, Hassan, Ashraf Sadik, and Toufic Mezher. "Present and Future Trend in the Production and Energy Consumption of Desalinated Water in GCC Countries." *International Journal of Thermal & Environmental Engineering* 5: 155 - 1 65. (accessed July 1, 2014).

29 Khaleej Times. " Desalination plants to run on renewable energy in Abu Dhabi." - Khaleej Times. http://www.khaleejtimes.com/kt-article-display-1.asp?xfile=data/nationgeneral/2014/May/nationgeneral_May48.xml§ion=nationgeneral (accessed July 2, 2014).

30 Khaleej Times. " Desalination plants to run on renewable energy in Abu Dhabi." - Khaleej Times. http://www.khaleejtimes.com/kt-article-display-1.asp?xfile=data/nationgeneral/2014/May/nationgeneral_May48.xml§ion=nationgeneral (accessed July 2, 2014).

31 International Energy Agency. "Fossil Fuel Consumption as a Proportion of Full Cost of Supply, 2012." . http://www.iea.org/subsidy/index.html pg. 22 (accessed July 1, 2014).

32 Brookings Institution . "These documents were prepared by the Brookings Institution for the 2013 Doha Carbon and Energy Forum. Doha Carbon and Energy Forum 2013." . http://www.dcef-qatar.org/app/media/7946 pg. 4(accessed June 1, 2014).

33 Brookings Institution . "These documents were prepared by the Brookings Institution for the 2013 Doha Carbon and Energy Forum. Doha Carbon and Energy Forum 2013." . http://www.dcef-qatar.org/app/media/7946 pg. 4(accessed June 1, 2014).

34 Brookings Institution . "These documents were prepared by the Brookings Institution for the 2013 Doha Carbon and Energy Forum. Doha Carbon and Energy Forum 2013." http://www.dcef-qatar.org/app/media/7946 pg. (accessed June 1, 2014).

35 "Qatar cautious on fuel subsidy: Al Sada." The Peninsula Qatar -. http://thepeninsulaqatar.com/news/qatar/216719/qatar-cautious-on-fuel-subsidy-al-sada (accessed July 2, 2014).

36 The Christian Science Monitor. "Fuel subsidies get scrutiny at Doha talks." The Christian Science Monitor. http://www.csmonitor.com/Environment/2012/1203/Fuel-subsidies-get-scrutiny-at-Doha-talks (accessed July 2, 2014).

37 "Renewable energy in Qatar." *Norton Rose Fulbright*: http://www.nortonrosefulbright.com/knowledge/publications/48660/renewable-energy-in-qatar. (accessed July 1, 2014).

38 British Petroleum. "BP Statistical Review of World Energy June 2014." . http://www.bp.com/content/dam/bp/pdf/Energy-economics/statistical-review-2014/BP-statistical-review-of-world-energy-2014-full-report.pdf (accessed July 1, 2014).

39 Production: 2003 – 949
Consumption: 2003 – 95
10% of Production was consumed domestically
Production: 2013 – 1995 (grew 110%)
Consumption: 2013 – 267 (grew 181%)
13.4% of production is consumed domestically

40 "Qatar Solar Energy." Qatar Solar Energy. http://www.qatarsolar-energy.com/ (accessed July 2, 2014).

41 Rao, Govinda, and May Al-Kuwari . "Assessment of Solar and Wind Energy Potential in Qatar." *02nd Joint Qatar-Japan Environment Symposium.* https://www.jccp.or.jp/international/conference/docs/14assessment-of-solar-and-wind-energy-potential-in.pdf. (accessed July 1, 2014).

42 MEED Insight. "Renewable and Alterative Energy." . http://www.meed.com/Journals/2012/03/19/b/w/f/Sample-chapter.pdf (accessed June 1, 2014).

43 "Qatar Solar Energy." Qatar Solar Energy. http://www.qatarsolar-energy.com/ (accessed July 2, 2014).

44 "QF currently produces 3.3MW of power from eight sites in Education City that have photovoltaic systems. These sites include Texas A&M University at Qatar, Carnegie Mellon University in Qatar, Weill Cornell Medical College in Qatar, Georgetown University School of Foreign Service in Qatar and four new solar car park shades." http://www.qatarsolar-energy.com/.

45 "Qatar Solar Energy." Qatar Solar Energy. http://www.qatarsolar-energy.com/ (accessed July 2, 2014).

46 "QF expands solar power production as part of renewable energy push - Doha News." Doha News. http://dohanews.co/qf-solar-initiative/ (accessed July 2, 2014).

47 "Sun shines on Qatar Foundation solar initiative." Gulf-Times. http://www.gulf-times.com/qatar/178/details/397974/sun-shines-on-qatar-foundation-solar-initiative (accessed July 2, 2014).

48 "Qatar Solar Energy." Qatar Solar Energy. http://www.qatarsolar-energy.com/ (accessed July 2, 2014).

49 "Qatar sheds light on its solar-power future." - Middle East. http://www.aljazeera.com/news/middleeast/2014/06/qatar-solar-power-future-201468121043472545.html (accessed July 2, 2014).

50 "QF expands solar power production as part of renewable energy push - Doha News." Doha News. http://dohanews.co/qf-solar-initiative/ (accessed July 2, 2014).

51 "Renewable energy in Qatar." *Norton Rose Fulbright*: http://www.nortonrosefulbright.com/knowledge/publications/48660/renewable-energy-in-qatar. (accessed July 1, 2014).

52 MEED Insight. "Renewable and Alterative Energy." . http://www.meed.com/Journals/2012/03/19/b/w/f/Sample-chapter.pdf (accessed June 1, 2014).

53 Rao, Govinda, and May Al-Kuwari . "Assessment of Solar and Wind Energy Potential in Qatar." *02nd Joint Qatar-Japan Environment Symposium.* https://www.jccp.or.jp/international/conference/docs/14assessment-of-solar-and-wind-energy-potential-in.pdf. (accessed July 1, 2014).

54 MEED Insight. "Renewable and Alterative Energy." http://www.meed.com/Journals/2012/03/19/b/w/f/Sample-chapter.pdf (accessed June 1, 2014).

55 "Renewable energy in Qatar." *Norton Rose Fulbright*: http://www.nortonrosefulbright.com/knowledge/publications/48660/renewable-energy-in-qatar. (accessed July 1, 2014).

56 MEED Insight. "Renewable and Alterative Energy." . http://www.meed.com/Journals/2012/03/19/b/w/f/Sample-chapter.pdf (accessed June 1, 2014).

57 Fath, Hassan, Ashraf Sadik, and Toufic Mezher. "Present and Future Trend in the Production and Energy Consumption of Desalinated Water in GCC Countries." *International Journal of Thermal & Environmental Engineering* 5: 155 - 1 65. (accessed July 1, 2014).

58 "Food security plan getting final touches." The Peninsula Qatar -. http://thepeninsulaqatar.com/news/qatar/243335/food-security-plan-getting-final-touches (accessed July 2, 2014).

59 MEED Insight. "Renewable and Alterative Energy." http://www.meed.com/Journals/2012/03/19/b/w/f/Sample-chapter.pdf (accessed June 1, 2014).

60 MEED Insight. "Renewable and Alterative Energy." http://www.meed.com/Journals/2012/03/19/b/w/f/Sample-chapter.pdf (accessed June 1, 2014).

61 "Energy subsidies hamper Middle East." Centre for Economics and Business Research. http://www.cebr.com/reports/energy-subsidies-hamper-middle-east/ (accessed July 2, 2014).

62 Brookings Institution . "These documents were prepared by the Brookings Institution for the 2013 Doha Carbon and Energy Forum. Doha Carbon and Energy Forum 2013." http://www.dcef-qatar.org/app/media/7946 (accessed June 1, 2014).

63 UNDP. "Energy Subsidies in the Arab World." http://www.undp.org/content/ahdr/en/home/resources/research-papers/energy-subsidies/ (accessed June 1, 2014).

64 Brookings Institution . "These documents were prepared by the Brookings Institution for the 2013 Doha Carbon and Energy Forum. Doha Carbon and Energy Forum 2013." . http://www.dcef-qatar.org/app/media/7946 (accessed June 1, 2014).

65 "U.A.E. looks to tap vast potential of wind power." UAE. http://www.uaeinteract.com/docs/U.A.E._looks_to_tap_vast_potential_of_wind_power/62304.htm (accessed July 2, 2014).

66 "GCC countries: Rising stars in solar and wind." Apricum. http://www.apricum-group.com/gcc-countries-rising-stars-solar-wind/ (accessed July 2, 2014).

67 MEED Insight. "Renewable and Alterative Energy." . http://www.meed.com/Journals/2012/03/19/b/w/f/Sample-chapter.pdf (accessed June 1, 2014).

68 "Qatar to boost renewable energy with reservoir rooftop solar panels - Doha News." Doha News. http://dohanews.co/qatar-to-boost-renewable-energy-with-reservoir-rooftop-solar-panels/ (accessed July 2, 2014).

69 "Dh1b Dubai solar park phase 2 tender opened by Dewa." Newsletter. http://gulfnews.com/business/general/dh1b-dubai-solar-park-phase-2-tender-opened-by-dewa-1.1319458 (accessed July 2, 2014).

References

Bakr, Amen. "Cheap UAE Gasoline: Citizens and Industry Square up." *Reuters*, May 30, 201. Accessed July 2, 2014. http://www.reuters.com/article/2012/05/30/us-emirates-petrol-idUSBRE84T0VW20120530.Bloomberg. "Dubai Denies Emirate Neighbors Fuel in Struggle to Pay Debt: Arab Credit." Bloomberg.com. http://www.bloomberg.com/news/2011-06-27/dubai-denies-neighbors-fuel-as-city-struggles-to-pay-off-debt-arab-credit.html (accessed July 2, 2014).

British Petroleum. "BP Statistical Review of World Energy June 2014." .http://www.bp.com/content/dam/bp/pdf/Energy-economics/statistical-review-2014/BP-statistical-review-of-world-energy-2014-full-report.pdf (accessed July 1, 2014).

Brookings Institution. "These documents were prepared by the Brookings Institution for the 2013 Doha Carbon and Energy Forum. Doha Carbon and Energy Forum 2013." . http://www.dcef-qatar.org/app/media/7946 (accessed June 1, 2014).

The Christian Science Monitor. "Fuel subsidies get scrutiny at Doha talks."The Christian Science Monitor. http://www.csmonitor.com/Environment/2012/1203/Fuel-subsidies-get-scrutiny-at-Doha-talks (accessed July 2, 2014).

Davis, Lucas. "The Economic Cost of Global Fuel Subsidies.". http://ei.haas.berkeley.edu/pdf/working_papers/WP247.pdf (accessed June 1, 2014).

"Desalination plants to run on renewable energy in Abu Dhabi." UAE. http://www.uaeinteract.com/docs/Desalination_plants_to_run_on_renewable_energy_in_Abu_Dhabi/61530.htm (accessed July 2, 2014).

Dh1b Dubai solar park phase 2 tender opened by Dewa." Newsletter. http://gulfnews.com/business/general/dh1b-dubai-solar-park-phase-2-tender-opened-by-dewa-1.1319458 (accessed July 2, 2014).

"Dubai in Top 3 Middle East Growth Markets for Solar Industry – Enpark & Mesia Supporting the Solar Industry." http://www.enpark.ae/media/press-releases/237-dubai-in-top-3-middle-east-growth-markets-for-solar-industry-enpark-a-mesia-supporting-the-solar-industry (accessed July 2, 2014).

Energy subsidies hamper Middle East." Centre for Economics and Business Research. http://www.cebr.com/reports/energy-subsidies-hamper-middle-east/ (accessed July 2, 2014).

"EnPark, Mesia support solar industry in Dubai." UAE. http://www.uaeinteract.com/docs/EnPark,_Mesia_support_solar_industry_in_Dubai/61722.htm (accessed July 2, 2014).

Fath, Hassan, Ashraf Sadik, and Toufic Mezher. "Present and Future Trend in the Production and Energy Consumption of Desalinated Water in GCC Countries." *International Journal of Thermal & Environmental Engineering* 5: 155 - 1 65. (accessed July 1, 2014).

"Food security plan getting final touches." The Peninsula Qatar -. http://thepeninsulaqatar.com/news/qatar/243335/food-security-plan-getting-final-touches (accessed July 2, 2014).

"GCC energy subsidies too high, ministers say." Latest News, Pictures, Video, Multimedia. http://m.gulfnews.com/business/oil-gas/gcc-energy-subsidies-too-high-ministers-say-1.1253614 (accessed July 2, 2014).

GCC countries: Rising stars in solar and wind." Apricum. http://www.apricum-group.com/gcc-countries-rising-stars-solar-wind/ (accessed July 2, 2014).

International Energy Agency. "Fossil Fuel Consumption as a Proportion of Full Cost of Supply, 2012." . http://www.iea.org/subsidy/index.html (accessed July 1, 2014).

Khaleej Times. " Desalination plants to run on renewable energy in Abu Dhabi." - Khaleej Times. http://www.khaleejtimes.com/kt-article-display-1.asp?xfile=data/nationgeneral/2014/May/nationgeneral_May48.xml§ion=nationgeneral (accessed July 2, 2014).

MEED Insight. "Renewable and Alterative Energy." . http://www.meed.com/Journals/2012/03/19/b/w/f/Sample-chapter.pdf (accessed June 1, 2014).

"Qatar to boost renewable energy with reservoir rooftop solar panels - Doha News." Doha News. http://dohanews.co/qatar-to-boost-renewable-energy-with-reservoir-rooftop-solar-panels/ (accessed July 2, 2014).

"Qatar cautious on fuel subsidy: Al Sada." The Peninsula Qatar -. http://thepeninsulaqatar.com/news/qatar/216719/qatar-cautious-on-fuel-subsidy-al-sada (accessed July 2, 2014).

"Qatar sheds light on its solar-power future." - Middle East. http://www.aljazeera.com/news/middleeast/2014/06/qatar-solar-power-future-201468121043472545.html (accessed July 2, 2014).

"Qatar Solar Energy." Qatar Solar Energy. http://www.qatarsolar-energy.com/ (accessed July 2, 2014).

"QF expands solar power production as part of renewable energy push - Doha News." Doha News. http://dohanews.co/qf-solar-initiative/ (accessed July 2, 2014).

"U.A.E. Oil Minister: Cut Energy Subsidies to Lower Domestic Consumption - Middle East Real Time - WSJ." Middle East Real Time RSS. http://blogs.wsj.com/middleeast/2014/01/21/u-a-e-oil-minister-cut-energy-subsidies-to-lower-domestic-consumption/ (accessed July 2, 2014).

"U.A.E. looks to tap vast potential of wind power." UAE. http://www.uaeinteract.com/docs/U.A.E._looks_to_tap_vast_potential_of_wind_power/62304.htm (accessed July 2, 2014).

"U.A.E. ranks 3rd among world nations in total concentrated solar power capacity." UAE. http://www.uaeinteract.com/docs/U.A.E._ranks_3rd_among_world_nations_in_total_concentrated_solar_power_capacity/62109.htm (accessed July 2, 2014).

UNDP. "Energy Subsidies in the Arab World." . http://www.undp.org/content/ahdr/en/home/resources/research-papers/energy-subsidies/ (accessed June 1, 2014).

Rao, Govinda, and May Al-Kuwari . "Assessment of Solar and Wind Energy Potential in Qatar." *02nd Joint Qatar-Japan Environment Symposium* https://www.jccp.or.jp/international/conference/docs/14assessment-of-solar-and-wind-energy-potential-in.pdf. (accessed July 1, 2014).

"Renewable energy in Qatar." *Norton Rose Fulbright*: http://www.nortonrosefulbright.com/knowledge/publications/48660/renewable-energy-in-qatar. (accessed July 1, 2014).

"Sun shines on Qatar Foundation solar initiative." Gulf-Times. http://www.gulf-times.com/qatar/178/details/397974/sun-shines-on-qatar-foundation-solar-initiative (accessed July 2, 2014).

"Water Desalination Using Renewable Energy: Technology Brief." *International Renewable Energy Agency* http://www.irena.org/DocumentDownloads/Publications/IRENA-ETSAP%20Tech%20Brief%20I12%20Water-Desalination.pdf. (accessed July 1, 2014).

About the Contributors

Ali Ahmad is a lecturer and researcher in Energy and Nuclear Policy at Princeton University's Woodrow Wilson School of Public and International Affairs. Ali's work covers energy economics, energy and climate change, nuclear security and nonproliferation and the introduction of nuclear power and renewables to new markets. A Physics graduate from the Lebanese University in Beirut, Ali Ahmad holds a PhD in Nuclear Engineering from Cambridge University.

Talmiz Ahmad joined the Indian Foreign Service in 1974, and served as Ambassador to Saudi Arabia, the UAE and Oman. Among his three authored books is *The Islamist Challenge in West Asia: Doctrinal and Political Competitions After the Arab Spring (2013).* Currently based in Dubai, he writes and lectures on Political Islam, West Asian politics, and energy security issues.

Hanan Albuflasa is assistent professor at the Physics Department of the University of Bahrain. She holds a BSc in Physics from the University of Bahrain (1993), a MSc in Renewable Energy System Technologies (CREST, 2004) and a PhD in Renewable Energy from the Loughborough University (2009). Albuflasa has a Diploma in Gemmology (GIA, UK, 1999) and a Diploma in Diamond Grading (Antwerp, Beligum, 2000). She is a member of International Solar Energy Society.

Mary B. Ayad is a lecturer at Macquarie Law School, she holds a BA in political science/pre law from the University of Colorado Boulder, an MA in social justice and international management from the School for International Training, an MA in international human rights law from the University of Malta and a doctorate in philosophy from the Macquarie Law School. Her research was awarded a prize by KPMG, the Mercantile Court of Arbitration of Spain and the Centre for European Studies, Madrid.

Gawdat Bahgat is professor of National Security Affairs at the National Defense University's Near East South Asia Center for Strategic Study. He is an Egyptian-born specialist in Middle Eastern policy, particularly Egypt, Iran, and the Gulf region. His areas of expertise include energy security, proliferation of weapons of mass destruction, counter-terrorism, Arab-Israeli conflict, North Africa, and American foreign policy in the Middle East. Bahgat's career blends scholarship with national security practicing. Before joining NESA in December 2009, he taught at different universities. Bahgat served as an advisor to several governments and oil companies.

Sujata Ashwarya Cheema is assistant professor at the Centre for West Asian Studies, Jamia Millia Islamia, New Delhi. Cheema was awarded the Government of Israel Graduate Fellowship (2002-03) and Visiting Research Fellowship (2011-12) to carry out research at Hebrew University of Jerusalem, Israel, and the Hermes Post-Doctoral Fellowship (2007) at Maison des Sciences de l'Homme, Paris. She was a Visiting Fellow at the MSH (2009) under the Indo-French Social Scientist Exchange Programme and a Visiting Researcher at the Department of Indian Studies, Faculty of World Studies, University of Tehran (2011). Cheema's publications include West Asia: Civil Society, Democracy and State (Ed) (2010).

Yoshikazu Kobayashi is senior economist at The Institute of Energy Economics, Japan (IEE Japan). Earlier he was an analyst at Tonen General Sekiyu, an ExxonMobil-affiliated company in Japan covering refinery operation planning and marine transportation. Kobayashi's research is on the world oil and gas market and energy security issues in Northeast Asia.

Østein Noreng has been professor (emeritus since 2013) of Petroleum Economics and Management at BI Norwegian School of Management since 1977 and from 1990 holder of the FINA Chair in petroleum economics and management. Since 1990 he has been Programme Director of the M.Sc. progamme in energy economics and managment at BI. Noreng has numerous teaching and research assignments at Norwegian and foreign universities and institutions. Further he has consultancies for Norwegian, foreign and international public and private organisations, including the governments of Norway, Denmark and Sweden, the US Department of State, the US Internal Revenue Service, the United Nations, the World Bank, the International Monetary Fund, the International Energy Agency, and several oil and gas companies.

M. V. Ramana is a physicist who works at the Nuclear Futures Laboratory and the Program on Science and Global Security, both at Princeton University, on the future of nuclear power in the context of climate change and nuclear disarmament. He is a member of the International Panel on Fissile Materials and the *Bulletin of the Atomic Scientists'* Science and Security Board. Ramana has written many papers and is the author of *The Power of Promise: Examining Nuclear Energy in India* (2012). Ramana is co-editor of *Prisoners of the Nuclear Dream* (2003) and author of *Bombing Bombay? Effects of Nuclear Weapons and a Case Study of a Hypothetical Explosion* (1999). He obtained his PhD in Physics from Boston University in 1994 and was a post-doctoral fellow at the Department of Physics, University of Toronto and the Center for International Studies, Massachusetts Institute of Technology. In 2014, Ramana received the Leo Szilard Award of the American Physical Society.

Dania Thafer is an independent researcher residing in Washington. She has a Master's degree in Political Economy from New York University. She has published several articles on the Gulf region including a monograph entitled *Iraqi Oil: A Looming Shift in Regional Dominance* with the Emirates Center for Strategic Studies.

Sara Vakhshouri is president of SVB Energy International, a Washington based strategic energy consulting firm that provides critical advice on the global energy market to private companies, governments, think tanks, investment banks, and media organizations. As an internationally recognized expert on the energy market and security, she has more than a decade of experience in global energy market studies, energy security and geopolitical risk. Vakhshouri has a PhD in energy security and Middle Eastern studies, an MA in business management (international marketing), and another MA in international relations. She has published articles in numerous journals including The Economist, Middle East Economic Survey, and Oil and Gas Journal. Vakhshouri is frequently quoted and appeared on Bloomberg, BBC, Financial Times, Reuters, Aljazeera, Energy Intelligence, Platts Energy TV and Voice of America on energy matters. She is the author of *The Marketing and Sale of Iranian Export Crude Oil since the Islamic Revolution.*